Q175 .B673 1992
Bohme, Gernot.
    Coping with science

DATE DUE

Library
Northwest Iowa Community College
Sheldon, Iowa 51201

DEMCO

Coping with Science

Coping with Science

Gernot Böhme

Westview Press
BOULDER • SAN FRANCISCO • OXFORD

All rights reserved. No part of this publication may be reproduced or transmitted in any form or by any means, electronic or mechanical, including photocopy, recording, or any information storage and retrieval system, without permission in writing from the publisher.

Copyright © 1992 by Westview Press, Inc.

Published in 1992 in the United States of America by Westview Press, Inc., 5500 Central Avenue, Boulder, Colorado 80301-2847, and in the United Kingdom by Westview Press, 36 Lonsdale Road, Summertown, Oxford OX2 7EW

Library of Congress Cataloging-in-Publication Data
Böhme, Gernot.
  Coping with science / Gernot Böhme.
    p.   cm.
  Includes bibliographical references and index.
  ISBN 0-8133-1237-X
  1. Science—Philosophy. 2. Science—Social aspects. I. Title.
Q175.G419   1992
303.48'3—dc20   91-42885
                                                                                          CIP

Printed and bound in the United States of America

The paper used in this publication meets the requirements of the American National Standard for Permanence of Paper for Printed Library Materials Z39.48-1984.

10   9   8   7   6   5   4   3   2   1

Contents

Acknowledgments		vii
1	The End of the Baconian Age	1
	Bacon's Conviction and Bacon's Program, 1 The 1970s and the Finalization of Science, 5 The Situation Today, 8 How Are Things to Proceed? 12 Notes, 15	
2	What Makes Us Feel We Need a Theory of Science?	19
	A Note on Terminology, 19 Notes, 26	
3	The Formation of the Scientific Object	29
	Notes, 39	
4	Can Science Reach Truth?	41
	Notes, 49	
5	Science and Other Types of Knowledge	51
	Science and Democracy, 52 Professionalization and Scientification, 54 Differences Between Types of Knowledge, 56 Summary, 61 Notes, 62 Bibliography, 62	
6	Science for War and Peace	65
	Useful Science, 65	

Historical Change as a Result of the Atomic Bomb? 67
Science in Resistance, 69
The Scientization of War and the Militarization
  of Science, 72
War Research as an Institution, 73
Is a Reversal of the Situation Possible? 74
Peace Research, 75
The Asymmetry of War and Peace, 76
The Idea of a Peaceful Science, 77
Notes, 78
Bibliography, 79

7 The Technological Civilization   81

The Relation of Man to Himself, 83
Social Relations, 86
Technical Civilization and Culture, 88
Notes, 90

8 The Knowledge-Structure of Society   91

Conclusion, 101
Notes, 102

9 An End to Progress?   103

The End of the Future? 103
The Concept of Progress, 105
Progress in Science, 106
An End to Progress by Science and Technology? 111
Notes, 113

About the Book and Author   115
Index   117

Acknowledgments

A first draft of this book was written during my stay at the University of Linköping in Sweden. My thanks go to the Department of Technology and Social Change at the TEMA Institute, which invited me to visit as a guest professor. I am also indebted to the Swedish Organisation of Universities and Colleges (UHÄ) at Stockholm as well as to the editors of the Graduate Faculty Philosophy Journal at the New School for Social Research in New York for their kind permission to reproduce Chapters 2, 3, 4, 7, 8, and 9. To Harry Redner of Monash University in Melbourne, Australia, I offer further gratitude. His steady insistence that I present my philosophy to the English-reading audience not only encouraged me to complete the book but also convinced the publisher to publish it. Finally, special acknowledgments go to John Farrell, who carefully translated Chapters 1 and 6, and to Spencer Carr, who oversaw the publishing process at Westview Press.

Gernot Böhme

1

The End of the Baconian Age

Bacon's Conviction and Bacon's Program

The usual practice is to name an epoch only when one has left it. It is only then that a feature, scarcely noticed beforehand, becomes conspicuous as a characteristic of the epoch; previously, such a feature would have been regarded as self-evident. The end of an epoch is thus characterized by the loss of self-evident truths. If we have occasion today to designate modern science as Bacon's epoch, then it is because we, in our relation to science, have lost something believed to be self-evident—namely, the basic conviction that scientific and technological progress is concomitantly human progress. This basic conviction was introduced into the world by Francis Bacon, and the beginning of the epoch of modern science was thus initiated. The conviction that scientific-technological progress is concomitantly human progress was not a self-evident truth before Bacon. On the contrary, the idea of progress was entirely foreign to the ancient world,[1] and human progress in the Christian Middle Ages was certainly not expected from either the worldly sciences; such progress could come, at best, from the religions of salvation. Bacon himself was well aware that he could not assume his conviction was shared by everyone. Rather, his entire literary activity was a singular propaganda campaign for this conviction.[2] He especially wanted to convince the political authorities (i.e., the king or queen ruling England at the time) of the "dignity and advancement of the sciences,"[3] and to persuade them to promote the sciences financially and organizationally. His program of great renewal (*instauratio magna*) appealed to those reformers[4] who wanted to extend the religious Reformation to a general reformation of the whole world.[5] What distinguished Bacon from the other reformers (among whom Oliver Cromwell and Johann Amos Comenius were representative) was the fact that he reckoned with neither political revolution nor educational reform; rather, he expected that the development of science and technology would secure the progress of the human race. Let us consider the famous Aphorism CXXIX from the first book of *Novum Organum:*

> For the benefits of discoveries may extend to the whole race of man, civil benefits only to particular places; the latter last not beyond a few ages, the former through all time. Moreover, the reformation of a state in civil matters is seldom brought in without violence and confusion; but discoveries carry blessings with them, and confer benefits without causing harm or sorrow to any.[6]

What constitutes Bacon's program? Briefly, it organizes science into an enterprise of inventions and institutionalizes it socially in such a way that its inventions are implemented for the benefit of man. Bacon's *Novum Organum* serves the first objective, and the second one is described in the utopia "New Atlantis."

The main intention of *Novum Organum* consists of making science into an innovative system. Since ancient times, it was indeed the primary objective of a scientist "to be learned," i.e., to know what was humanly possible to know.[7] The idea of the learned person presupposed a complete canon of knowledge, or at least one that could be completed. The accompanying tool of knowledge was the logic founded by Aristotle. It was an instrument with which one could move about systematically (i.e., argumentatively) in the system of knowledge. In his *Novum Organum,* however, Bacon opposed Aristotle's logic with a logic of invention. He called for an instrument with which one could make discoveries methodically. This method consists of the experimental process and induction. Aware of several inventions that had fundamentally changed the world, such as gun powder, the compass, and the printing process, Bacon took great pride in an invention on the meta-level, as it were—namely, the invention of a method to make inventions: "How much higher a thing to discover that by means of which all things else shall be discovered with ease!"[8]

This meta-level invention resulted in an epoch-making change for the system of human knowledge. From that point forward, knowledge was no longer a complete canon, nor one that could be completed; rather, it became the prototype of a continually self-renewing system. After Bacon, the main objective of the scientist was no longer to be a learned person but, instead, to be someone who has made or can make a "contribution"—that is, a contribution to the body of collective knowledge, which is continually being enlarged and renewed. The self-confidence of this new type of scientist created by Bacon was aptly expressed centuries later by Max Weber in his lecture entitled "Science as a Vocation": "In science, each of us knows that what he has accomplished will be antiquated in ten, twenty, fifty years. That is the fate to which science is subjected; it is the very *meaning* of scientific work."[9] Science as research, i.e., as a collective system organized around innovation: This was Bacon's demand, and that is what science became after Bacon.

The social integration of science and its organization around the principle of usefulness was mentioned as the second aim of the Baconian program. In Bacon's utopian state "New Atlantis," science is organized as "Salomon's House." Bacon conceptualized this institution as a second authority that, along with (and yet independent from) the political authority, would look after the public good. Salomon's House would operate a large number of research laboratories in which science is organized not in terms of disciplines but according to social use and fields of application. Thus, for instance, it would provide climate laboratories, labs for studying animal husbandry, optical workshops, and so on. The collection and distribution of knowledge is organized in Salomon's House, as is the popularization of knowledge and, conversely, the identification of knowledge gaps in the social field of application. In addition, Salomon's House would offer a series of direct social services, such as the forecasting of bad weather and earthquakes, floods, and possible famines. It also would organize the public health service through state officials.

Salomon's House was the concrete realization of Bacon's conviction that human conditions can be improved primarily through the development of science and technology. In this context, he saw the materialization of the biblical command "Subdue the Earth." The domination of the human species over all of nature (*Novum Organum,* Aphorism CXXIX) is to be achieved only through development of our knowledge of nature, for (to quote Aphorism III) "Human knowledge and human power meet in one; for where the cause is not known the effect cannot be produced. Nature to be commanded must be obeyed; and that which in contemplation is as the cause is in operation as the rule." Here, Bacon suggests that man's domination of nature consists not in his instigation of something against nature (recall antiquity's comprehension of mechanics)[10] but in his being able to use it for his own purposes by means of a precise knowledge of natural causes. Required, then, is the transformation of causal knowledge into rules for action, i.e., into technical procedures. Now, the crucial point here is that Bacon, in his programmatic work "New Atlantis," did not stop at this quasi-epistemological insight but, rather, called for a social institution that would perform the mediation of knowledge of nature and social needs. He thereby projected a society in which scientific knowledge itself is a social authority and in which science as research constitutes an important part of public life. This society we now have.

The Baconian program has been fulfilled to a degree Bacon could not have imagined. But has his conviction been borne out? Can scientific-technological progress be equated with human and social progress? Today, the answer would have to be "no." This negative response is something new, a deep sobering effect that demands a fundamental revision of our relation to science. Along the path toward fulfillment of the Baconian program, this question was for some time answered in the affirmative. But now we must admit that, so long as the implementation of the program itself was still incomplete (particularly in the nineteenth century, which designated itself the century of natural science), great hopes were placed in science and technology.[11] Socialism, too, augmented the Baconian program with the idea of an organization of the whole of society on a scientific basis.[12] A "defetishization of science"[13] is especially necessary in this connection, given the breakdown of real socialism.

This sobering effect, under further investigation here, allows us to pose the question today as to whether Bacon had no doubt whatsoever about the identification of scientific-technological progress with human progress. This question must be answered in the negative, although one does find in Bacon's model the rudiments of a distinction between neutral knowledge and morally responsible application—a distinction that has long served as a defense for science in the twentieth century. Note that paragraph CXXIX of *Novum Organum* closes with the following passage:

> Lastly, if the debasement of arts and sciences to purposes of wickedness, luxury, and the like, be made a ground of objection, let no one be moved thereby. For the same may be said of all earthly goods: of wit, courage, strength, beauty, wealth, light itself, and the rest. Only let the human race recover that right over nature which belongs to it by divine bequest, and let power be given it; the exercise thereof will be governed by sound reason and true religion.

The implication is that Bacon's confidence in reason and religion does not permit even the thought to arise that there could be a deep ambivalence about scientific-technological development. A further point that not only commands our attention today but also raises questions as to the consistency of Bacon's program and conviction is the isolation of "New Atlantis" from the rest of the world. (All members of Salomon's House were obliged to observe secrecy as a matter of principle.) Although Bacon quite plainly viewed the development of science and technology as a concern of humanity, he placed his utopian vision of a concretely useful and socially institutionalized science within the framework of nation-state egoism. Nor is this simply a question of presentation, to be ascribed to the literary form "utopia"; on the contrary, for Bacon one of the most important dimensions of scientific-technological development in terms of use was war. And *this* use can never be thought of in terms of humanity as a whole; rather, this use is possible only when one nation can avail itself of it or at least has a scientific-technological advantage over others. Here we see an obvious inconsistency in the Baconian conception of useful science; and, given this inconsistency, the modern scientization of war is one of the main objections to the Baconian program as a whole.

The 1970s and the Finalization of Science

The last renaissance of the Baconian program occurred recently, during the late 1960s and early 1970s. At this time, for various reasons and from different sides, great hopes were again placed in science. Fulfillment of the Baconian program resulted when science achieved the rank of an important sector of society. By socialist societies—which in any case regarded their well-being as based on a scientific organization of all of society—this phase was understood as the transition of science to its role as a leading productive force.[14] Analogously, the theory of postindustrial society that developed in the West held that theoretical knowledge has developed into an "axle" around which the new society will revolve.[15] Although this development was not so emphatically welcomed in the West (as it was in the East by the Richta study group, for example), the trend was clear here, too: From science, one expected a definitive satisfaction of basic human needs—that is, a multidimensional development of man and, especially, of his creativity.[16] During the 1970s, in all industrialized nations, the development of science and technology was studied intensely, with an emphasis on optimal organization and orientation of the field according to social needs. Responsible for this extraordinary appreciation of science were, among other factors, the achievement of space research (and the competition in this area), the initial successes of the so-called green revolution, and the extraordinary hopes placed in the peaceful use of nuclear energy. A wide-scale promotion of science set in, followed by an enormous expansion of universities and polytechnics; and science and technology policy emerged as a special concern of government administration. Even the student movement represented an expression of the special social appreciation of science. Indeed, as the students of the 1960s and 1970s perceived themselves to be the bearers of a future social development, they were able to lay claim to an independent political role.

As someone who at the time was deeply involved in the public debate about the social role of science and technology, I may perhaps be permitted a personal retrospection. The increased appreciation experienced by science then, the special political and social interest directed to it, led naturally enough to a scientific concern with science, to "science research" (as it is called today); its topic was science's societal effect. In this context there followed extensive debates about the priorities of research policy; our ability to control scientific development was argued at length, and the organizational form of science in terms of its effectiveness was investigated with the intention of overcoming traditional hierarchical structures. Finally, there was the question of the "relevance" of research itself, i.e., the question as to whether this potential would be correctly developed and for the right purposes—the potential from which the solution to almost all human problems was expected.

Included in this context is the research of the group "Alternatives of Science," to which I belonged as a member of the Starnberg *Max-Planck-Institut zur Erforschung der Lebensbedingungen der wissenschaftlich-technischen Welt* (Institute for the Research of Living Conditions in the Scientific-Technological World). The group's collaborators were strongly motivated by the "relevance of research" claim raised by the student movement. Through their work, they hoped to contribute to this relevance; they also wanted to remove the barriers set against a social orientation of science by conservative members of the scientific community. The objective was to exhibit the inner flexibility of research, the alternative latitude in the further development of science, and thereby to contribute to the possibility that this latitude would be used. The theory of the "finalization of science"[17] grew out of this research. Our point of departure was Thomas Kuhn's theory of the development of science.[18] Kuhn devised a two-phase model for the development of science—that is, the transition from revolutionary to normal science and vice versa. Normal science is dominated by exemplary and comprehensive performance in one area to such an extent that everyday science can progress as a kind of puzzle solving. However, because of anomalies, inconsistencies, and unexplained effects, a paradigm at times reaches the limits of its capabilities, at which point the phase of normal science is replaced by a revolutionary phase in which a new paradigm is strived for. Now, according to Kuhn, science could continue to roll on in an endless exchange between normal and revolutionary science. Prompted by the claim advanced by W. Heisenberg and C. F. von Weizsäcker as to the existence of "closed theories,"[19] we showed that old theories (e.g., Newtonian mechanics) replaced by new ones in revolutionary phases (e.g., Einstein's theory of relativity) are not rendered obsolete as a result; rather, they often continue to be valid in an object area that, though perhaps restricted, is nevertheless all the more precisely defined. Moreover, these old theories frequently form the basis for application-oriented developments.

The fact that old theories often become classic theories led us to introduce a three-phase model of the development of science. During the first phase of development of a new discipline, there is no paradigm available. In this phase, the manner of thematizing the subject matter is determined above all by the choice of measurement methods and the development of concepts. At this point, therefore, the science is still largely open to external influences. The second phase is characterized by various competing theory models. This phase, which we termed "paradigmatic," is largely autonomous because it entails an inner-scientific clarification of the consistency and range of theories, and largely acquires its problem complex from the competition between theories. Finally, however, this phase is closed when one theory becomes dominant. The theory is then designated mature insofar as it solves in principle all the problems of an object area. As Heisenberg and von Weizsäcker had already discovered, such a theory can no longer be improved by minor alterations—and major alterations lead into a revolutionary phase in Kuhn's sense. The lasting validity of a mature theory for an object area does not lead to a termination of development, even without revolutions. Rather, the finalization phase follows. This is the phase during which a mature theory is further developed in terms of various application fields. That the matter is one of further theoretical development at all is precisely due to the fact that the basic theory grasps its object area in principle only. To make the basic theory applicable to concrete individual cases, additional concepts and mathematical methods ("application concepts," as we called them) are necessary. The specialization of the basic theory arising from this further development is determined by targeted social purposes.

By examining science in terms of its development, we showed that there are reasons in science itself that, depending on the phase, open it to or close it off from external influences. The social orientation of science is therefore legitimate—indeed, sometimes even required, at least in certain phases.

These theses, which were validated by numerous case studies, triggered an extremely intense public debate at the time.[20] This furor would be difficult to understand from today's perspective. Far from being ideological, our investigations were quite dispassionate. And what we were claiming with respect to the social orientation of scientific development was a phenomenon obvious to everyone, one that motivated the public interest in science. However, people were not ready to accept what we had revealed through our investigations: that the social orientation of science affects its inner structure, penetrates down to the concepts themselves, and, in the final analysis, is responsible for the kind of knowledge we now possess. Accordingly, scientific knowledge is not a neutral instrument. A further conclusion is that the distinction between basic research and application, or that between science and technology, cannot be maintained as a rule. Accordingly, the barriers erected by conservative minds to ward off science's social responsibility must fall.

The Situation Today

Looking back today, I believe that the disputes surrounding the theory of the finalization of science were, without exception, ideological ones. The central issue was not so much the reality of science as the determination of what had to be thought and said about science in public. This is an important issue, of course; and we must keep in mind that what is on today's agenda is a change in the attitude toward science and in the self-awareness of scientists.

With the approach of the 1970s, one became aware that the Baconian program was indeed fulfilled. Science, now a system of research through and through, had been completely institutionalized as an innovative system at the societal level. The furor that resulted was generated by the academic resistance to this process. By contrast, our own claim concerning science's "social relevance" was forcing the door open; we were merely expressing what had already long been the case. By way of qualification I must add, however, that science's social orientation was not motivated by purely desirable aims; on the contrary, it had been and was still being institutionalized at the societal level mainly as an instrument of military and economic competition.[21]

What is the situation today? One thing is clear: The euphoria of the 1970s has disappeared—along with the heated character of the discussions about science. Science and technology policy is still an important area of politics, and the public and private resources invested in the promotion of science are enormous. However, the investment of social resources in promoting science and technological development is no longer being legitimated by promises of salvation. And no longer invoked is the basic conviction, characteristic of the Baconian age, that scientific and technological progress is at the same time human progress. Instead, arguments are bluntly presented from the perspective of economic and military competition: If we do not promote science and technological development intensively, we will fall behind in the international competition. Above all (and this may prove to be the most important change because it is not a question of opinions but one of facts), the overcapacity in the area of scientific research and technological development is being noted for the first time. Granted, worldwide military disarmament has not yet taken hold in the corresponding scientific sector (i.e., military research and development). But a reduction in R&D capacity will be necessary if the policy of détente is to take hold of this sector as well. The conversion of scientific capacity from war research to peace research, which has been preached to scientists in order to render more palatable the thought of a demilitarization of science, has proven illusory. For a peaceful world and thus also a peaceful science, there are simply too many scientists and engineers. And the fact that one can no longer invoke humanitarian aims to legitimate R&D expenditures indicates that the Baconian conviction is no longer a universal supposition on permanent stand-by.

The doubts about the harmony between scientific-technological and human progress have three main causes. The first concerns the relationship between science and war. Science has indeed been discredited by the fact that by far the largest part of the expenditures for scientific-technological manpower has hitherto been invested in the military sector.[22] Even if one considers the remaining "blessings" of scientific-technological development to be beyond doubt, the destruction potential given to man by science is nevertheless far greater than all of these blessings taken together. Moreover, the relationship between the particular use and the universal use of science (a relationship that probably represents an inconsistency in Bacon's comprehension of science's usefulness) has developed into a hopeless dialectic. Science's usefulness for war (clearly the use for which most funds have been invested) is always merely a particular and temporary usefulness. Hence science is useful only so long as it furnishes one party with a temporary advantage. For humanity as a whole, then, science is basically harmful in the realm of war.

The second reason as to why, for us, the Baconian belief has been shaken was concealed in Bacon's program by his confidence in reason and religion. Because the moral forces of the twentieth century have not been sufficient to compel the use of science and technology for the good, the question arises as to whether we are dealing with a value-neutral instrument at all. At least Bacon had assumed that the effects of science and technology are in every case good when they are used for good intentions. However, it is precisely in this context that serious doubts, and matters of principle, emerge. Given that every technological intervention—in nature, in the human body, in society—includes not only effects but side effects too, and as these effects and side effects cannot be distinguished in the matter itself (i.e., irrespective of human intentions), one must expect that with every useful employment of science and technology, harm can be caused. This insight has been thematized more generally as the dialectic of progress. Progress entails regress, too;

and striving for human progress by way of science and technology (i.e., by means of rationality) exacts a specific human price. This price, more generally, is the price of modernity: one-dimensionality, the loss of immediacy, instrumentalization of the body, the desolation of the lifeworld's knowledge forms and competencies, affective disorientation, and chaotic emotions.[23] These dark sides of rationality, which appear on the ideological level merely as points in the critique of culture, are in truth very concrete problems and causes of suffering. Thus, for instance, the instrumentalization of the body takes its revenge in the form of psychosomatic suffering. The improvements in midwifery resulting from rationalization and hospitalization are dearly paid for by the difficulty in integrating childbearing and the child into one's own life-context.[24] Thus, too, the bureaucratization of administration leads to a loss of humaneness in social intercourse. These civilization problems are examples, at a higher level, of the same dialectic revealed by intervention in nature: Fertilization leads in the long term to barren soil, regulation of rivers and lakes leads to a lowering of the ground-water level and to the necessity for irrigation, the use of antibiotics leads to the spread of allergies, and so on.

The third cause responsible for unsettling the Baconian belief in human progress via science and technology has gained specifically political significance with the failure of scientific socialism. In all industrialized nations, the influence of scientists on politics has increased both with the growth in complexity of decisions and with the expansion of state regulation. This influence is mediated by political advising on the part of scientists and the scientifically trained civil service; under the rubric of "technocracy" it has also led in the West to the tendency to replace politics with science. In the countries of "real socialism," however, this tendency has been significantly reinforced and legitimated by the idea that societies develop according to laws of history. Here, the Baconian conviction had taken the form of a belief in being able to transform human society into an authentically humane one by means of rational planning and law-like development. Repression, servility, loss of self-initiative, and, finally, economic collapse were the result. The consequences for world history of the failure of this social experiment are not yet clear. After all, we are dealing here with the failure of an *idea* as well—the idea of a scientifically organized society. This failed idea will also have consequences for the appraisal of the social significance of science in the West. Even today it can be said that scientific knowledge—even though it still ranks highest in the knowledge hierarchy—is no longer regarded as the *decisive* knowledge form. Rather, it is well known that the political disputes and social antagonisms for whose resolution one engages scientists are repeated in the scientist's statements. This antagonistic situation was particularly evident in the disputes about nuclear power plants.[25]

The situation today is thus characterized by considerable disillusionment with respect to Bacon's hopes relating to science and technology. Precisely in those countries where the Baconian program of a social institutionalization of science has been fully realized, science is no longer being legitimated as a means for the advancement of humanity. Rather, it is at most a necessary instrument for "survival"—whether survival in the major catastrophes into which humanity has maneuvered itself or simply survival in the competitive world. Indeed, the major problems of humanity, such as overpopulation, the destruction of nature (humanity's basis for life), and the exhaustion of natural resources, were created by scientific-technological progress in the first place. Although this situation has led to an antiscience movement, it has certainly not brought about either a dismantling of science or a diminishing of the willingness to expend enormous public funds on science. What we face today, then, is the realization of the Baconian program without the Baconian belief. Science is an important sector of social life, and as a method for problem solving it shapes almost all realms of social life. Universities, independent research institutes, large research installations, and state and industrial research centers combine to form an enormous social institution. This is because an appreciable part of social resources is invested in science, and a major part of social labor is scientific labor. Moreover, in terms of social position and numbers, scientists and engineers represent at least a potential power factor. Although the "intelligentsia" is not to be regarded as a new dominating class, it is nevertheless a group with privileges and its own interests, a group that could form into a class under conditions of scarcity.[26] But Salomon's House is no utopia. We live in a society where science and technology constitute a major social institution and thus wield real social power.

Our society is also characterized by scientization. Science is not an authority extraneous to society, from which social processes are to be recognized and steered; rather, it is embedded in the social structures themselves. The social infrastructure,[27] which today consists of communication networks and data processing systems, is indeed a scientific and technological entity. And social life, consisting of public-opinion formation, democracy, adjudication, and administration, takes place within the framework of the infrastructure instituted by science and technology. Even individual spheres of action are scientized in the sense that they build on scientific technology, and action implementations are organized according to scientific rationality. This point is most clearly illustrated by the scientization of war. From reconnaissance through military equipment to combat management, modern war is a scientifically organized process. The same is true of "natural" processes such as giving birth, when it occurs within the framework of modern obstetrics. Industrial labor in the industrialized nations is also a scientific-technological process, as are administration and politics. Although political decisions have not been scientized in the sense of having been technocratically replaced by scientific ones, the implementation of administrative and political decisions would be inconceivable in the absence of the scientific-technological infrastructure, which encompasses such factors as data collection, planning concepts, and of course, scientific theories.

If one takes these two points together—science as an institution and the scientization of social spheres—it is impossible to consider science and technology as being more or less independent variables and to ask whether it promotes social and human progress. Rather, science and technology today belong to our lifestyle. The question, then, is not whether science and technology make human life more humane; rather, it is what humanity means at all under the conditions of a life-form that is already largely shaped by science and technology. We unavoidably live in a technological civilization; but we have not yet come to terms with the fringe conditions of this life-form: "Coping with science" is really the key issue here.

How Are Things to Proceed?

How are things to proceed without these Baconian hopes? The social significance of science and technology is greater than ever, and the development of science far more dynamic—and this without any promises of paradise or of a solution to the mystery of the universe. Defensive and legitimizing strategies are the immediate consequences of this situation. Science and its institutions must engage in public relations. Popular science, science journalism, and the effective stage-managing of science are required. Using full-page glossy advertisements, industries dependent on science (such as the chemical industry) struggle against the fact that *chemical* has become a pejorative word in everyday language. Current research in science and technology, promoted and stage-managed by the state, has objectives that are completely different from those of the 1970s: It now serves the purpose of "cushioning" scientific and technological progress. Because the humanitarian use of this progress has become questionable, research now entails the "acceptance" of technological progress in society. In other words, the progress must be made "socially and environmentally acceptable," if possible. Admittedly, these strategies also require that one come to terms with science. And far be it from me to contest their necessity. However, I do criticize the fact that these strategies are pursued impetuously by the scientific community as methods of defense and repair. Accordingly, they fail to reflect anew our relationship to science.

Such a reflection would have to take the fact of science as its point of departure. Like writing and the wheel, science and technology belong to the *condition humaine* today. Yet they have not been integrated into the context of humanity. In this connection Günter Anders points to "the antiquatedness of man."[28] However, I am not of the opinion that man with his moral capabilities has fallen behind his technological capabilities and that the task of coming to terms with science would now require him to move morally to the level of scientific and technological development. To rely on a higher moral development of humanity would mean robbing Peter to pay Paul—a strategy that shares with the Baconian ideology of science the fetishization of progress. No, what matters is to separate the question of what is human and what is not from the idea of progress and thus also from the idea of scientific progress. *First,* a new reflection of our relationship to science must pose the question as to what science is and what it is not.[29] The expectations placed on science must be reduced if we are to maintain its undoubted value. We therefore have to clarify both what happens to an object or to an object area when it is scientifically thematized and what is special about the treatment of a problem with scientific methods. And in order to do this we have to regard science once more as a possible knowledge form on a wide spectrum of knowledge forms. Then we will understand that science is a knowledge form with its own particular structure and that this knowledge form is meaningful only in specific contexts with specific action intentions. Locating science within a wider spectrum of scientific knowledge forms requires that we re-estimate nonscientific forms of knowledge.

Second, the relationship between science and education must be thematized. The inevitability of science can also be characterized by the fact that our state of culture is technological civilization. Technological civilization is indeed a fact, but one that has "organically" come upon us. The old ideal of a classical education, which was dominant in Europe up to and throughout the nineteenth century and was determined by the linguistic and human sciences, has been suppressed step by step in the twentieth century. What has taken its place—namely, natural science and technology—has thus far been comprehended only as instrumental knowledge. In their educational content (i.e., with respect to the extent that these knowledge forms educate man), science and technology have not yet become transparent. Indeed, following the publication of C. F. Snow's *The Two Cultures,*[30] the possible dichotomization of Western culture—into a scientific-technological culture and a historical-literary culture—was heatedly debated. Yet the actual composition of these cultures was not investigated. On the one hand, of course, we might assume that scientists are socialized into specific types of thinking during their professional training. But whether such socialization is equivalent to education and not merely instrumentalization according to specific professions and roles is unclear. On the other hand, A. W. Gouldner claimed in his book *The Future of Intellectuals and the Rise of the New Class* that all intellectuals, from arts-page writers to engineers, had a common culture—namely, the culture of critical discourse.[31]

The cultural significance of science must still be made transparent. And science itself must then be run as a culture-creating enterprise. Such an enterprise can come about only if, *third,* science is liberated from today's dominant objective: usefulness. But one cannot bring about this liberation by reverting to the old distinction between basic research and application. As has been shown, basic research today is legitimated in terms of the foundations for application; and, ultimately, its content is affected by intended applications (finalization). One must therefore pursue science for the sake of knowledge and as an orientation in the existing world. Toward this end, a concern with deep-seated foundations is by no means necessary. As an example, consider the research on the colors of butterfly wings. In 1979, Sam Allison described such research as a threat to state science policy, which wanted to make science a state secret.[32] Today, however, it must be noted that natural science has moved into realms that provide us with knowledge about nature but not with knowledge about how to manipulate nature. I am thinking about the study of fractals and chaos. This research, which does indeed fundamentally change our picture of nature, demonstrates that man's potential for manipulating nature is limited. Here, the issue is one of dissolving the close connection between knowledge and power,' between scientific knowledge and production, which was asserted in Aphorism III of Bacon's *Novum Organum*. Many approaches to the sciences, such as Goethe's theory of colors,[33] were not further developed in our culture because they could not be converted into technology. Yet the ambivalence of the use orientation of scientific knowledge could lead to a new openness vis-à-vis other forms of science.

Our relationship to nature can be mentioned as the *fourth* and final point concerning the new reflection of science that has become necessary. Scientific knowledge always constitutes a relationship to nature, either to outer nature or to "the nature that we are ourselves," that is, the body.[34] Within the framework of the Baconian program, this relationship to nature was shaped as a relationship of dominance. What was sought and recognized as scientific knowledge was specifically the knowledge that would extend man's domination of nature. In the beginning, this program of domination seemed plausible because it promised to free man from labor and worry; and this promise has largely been fulfilled. In the meantime, however, the dialectic of progress has asserted itself, too: Scientific-technological progress has led to an overburdening of man, who now finds it *necessary* to manipulate. This dialectic has brought to light what has been overlooked in the concept of nature domination—namely, the fact that dominated nature must "play along." In this connection, by the way, the reader should keep in mind that Bacon himself had more insight than did those successors who acknowledged his program. As already noted, Bacon himself pointed out that nature can be dominated only by obeying it. However, mankind subsequently overlooked the fact that, in all aspects of the domination of nature, man is almost entirely dependent on nature's own activity. The progress of nature domination has suppressed and made ineffective nature's own activity such that the necessity for man to maintain these processes, to steer and reproduce systems by means of conscious manipulation, has continually increased. Today, a large part of human labor, of the use of science and technology, and of the need for their further development must be ascribed to the fact that nature no longer relieves man of labor; rather, man burdens himself with labor of a type that nature did by itself in the past. Thus, it can be said that the project of nature domination has become obsolete. In short, knowledge today must aim not at the domination of nature but at the cooperative interplay between nature's own activity and human use.

With the end of the Baconian age, Bacon's program has been realized, but his hopes have not been fulfilled. We have to live with science because it is tied in with the conditions of our very existence. Expectations of salvation will no longer be attached to it, however. This situation will also necessitate a concrete dismantling of scientific-technological capacities—but, above all, a dismantling of the exaggerated expectations associated with scientific-technological development. This dismantling may be painful, and it will likely generate resistance as well. At the same time, however, it represents the liberation of the mind from a use-oriented knowledge form, and the liberation of man himself from a one-dimensional orientation toward rationality. It also represents the liberation of the individual human being from domination by experts. Knowledge itself must be liberated from its position as a service industry. It should delight once again; it should enhance the education of man and his everyday competence. And, finally, it should contribute to wisdom.

Notes

This chapter has been translated into English by John Farrell.

1. On this point, see W. Krohn, "Die 'Neue Wissenschaft' der Renaissance," in G. Böhme, W. v.d. Daele, and W. Krohn, *Experimentelle Philosophie. Ursprünge autonomer Wissenschaftsentwicklung* (Frankfurt: Suhrkamp, 1977).

2. W. Krohn, *Francis Bacon* (Munich: C. H. Beck, 1987).
3. Francis Bacon, *The Advancement of Learning*, 1603.
4. C. Webster, *The Great Instauration: Science, Medicine and Reform 1626–1660* (New York: Holms and Meier, 1976).
5. *Allgemeine und General Reformation der Gantzen Weiten Welt* (Universal and General Reformation of the Whole Wide World) is the title of a book by the Rosicrucian Johann Valentin Andreae, published in 1614.
6. Francis Bacon, *The New Organon and Related Writings*, edited, with an introduction, by Fulton H. Anderson (New York: Liberal Arts Press, 1960), p. 117.
7. F. Znaniecki, *The Social Role of the Man of Knowledge* (New York: Octagon Books, 1975).
8. Bacon, *op. cit.*, Book I, Aphorism CXXIX, p. 119.
9. Max Weber, "Science as a Vocation," in H. H. Gerth and C. Wright Mills (eds.), *From Max Weber: Essays in Sociology* (London: Routledge & Kegan Paul, 1970), p. 138.
10. Mechanics as μηχανή (mechané or trick).
11. Speaking as a representative for many others, H. von Helmholtz made the following statement in his "Academic Speech," held at the inauguration to the Prorectorate in Heidelberg in 1862: "In point of fact, the men of science do form a kind of organized army. For the betterment of the whole nation, and almost always on its behalf and at its cost, they try to enlarge the knowledge which can serve both the increase in industry, wealth and the beauty of life as well as the improvement of the political organization and moral development of individuals." *Vorträge und Reden* (Braunschweig: Vieweg, 1903), p. 181.
12. One thinks of Lenin's maxim: "Electrification and socialism."
13. W. Fedotova (Academy of the Sciences of USSR), "The Social Scientist's Responsibility and the Expectations Associated with Science," lecture held in Kassel, 1990.
14. Richta-Report, *Politische Ökonomie des 20. Jahrhunderts* (Civilization at the Crossroads) (Frankfurt: Makol, 1971).
15. D. Bell, *The Coming of Post-Industrial Society: A Venture in Social Forecasting* (New York: Basic Books, 1973).
16. For a related discussion, see H. Marcuse, *Eros and Civilisation: A Philosophical Inquiry into Freud* (Boston: Beacon, 1974).
17. G. Böhme, W. v.d. Daele, and W. Krohn, "Finalization in Science," *Social Science Information,* 15 (1976), pp. 307–330. The main research results of the group "Alternatives of Science" are to be found in G. Böhme, W. v.d. Daele, and K. Krohn, *Experimentelle Philosophie. Ursprünge autonomer Wissenschaftsentwicklung* (Frankfurt: Suhrkamp, 1977); and G. Böhme et al. (eds.), *Finalization in Science: The Social Orientation of Scientific Progress* (Boston: Reidel, 1983). My colleague W. Krohn has written a different retrospective on the theory of finalization: "Finalisierung der Wissenschaft—Retrospektive und Perspektive," *Arbeitsblätter zur Wissenschaftsgeschichte,* 22, Halle, Martin-Luther-Universität, 1989.
18. Thomas Kuhn, *The Structure of Scientific Revolutions* (Chicago: University of Chicago Press, 1962).

19. W. Heisenberg, "Der Begriff 'abgeschlossene Theorie' in der modernen Naturwissenschaft," in W. Heisenberg, *Schritte über Grenzen* (Munich: Piper, 1971), pp. 87–94; C. F. von Weizsäcker, *Die Einheit der Natur* (Munich: Hanser, 1972), p. 193. G. Böhme, "On the Possibility of Closed Theories," *Studies in the History of Science,* 11, 2 (1980), pp. 163–172.

20. A bibliography can be found in Böhme et al. (eds.), *op. cit.;* and in W. Schäfer, *Die unvertraute Moderne* (Frankfurt: Fischer, 1985).

21. This has been the case until recently. Cf. Rilling et al., "Dossier Forschungspolitik," *Forum Wissenschaft,* No. 2 (1988): I–XIX.

22. For recent figures on war research, see Mary Acland-Hood, "Military Research and Development Expenditure," in *Sipri-Yearbook,* 1985; on the situation in West Germany, see Dossier Forschungspolitik, *op. cit.*

23. The classic work on this is Max Horkheimer and Theodor W. Adorno's *Dialectic of Enlightenment* (New York: Social Studies Association, 1944). Our critique of reason has continued along its lines; in Harmut and Gernot Böhme, *Das Andere der Vernunft. Zur Entwicklung von Rationalitätsstrukturen am Beispiel Kants,* 2nd ed. (Frankfurt: Suhrkamp, 1985).

24. G. Böhme, "Midwifery as Science: An Essay on the Relation Between Scientific and Everyday Knowledge." In N. Stehr and V. Meja (eds.), *Society and Knowledge: Contemporary Perspectives on the Sociology of Knowledge* (New Brunswick and London: Transaction Books, 1984), pp. 365–385.

25. H. Novotny, *Kernenergie. Gefahr oder Notwendigkeit* (Frankfurt: Suhrkamp, 1979).

26. A. W. Gouldner, *The Future of Intellectuals and the Rise of the New Class* (New York: Seaburg Press, 1979).

27. G. Böhme, "The Techno-Structure of Society," *Thesis Eleven,* 23 (1985), pp. 104–116.

28. G. Anders, *Die Antiquiertheit des Menschen,* 2 vols., 4th and 7th eds. (Munich: Beck, 1987).

29. G. Böhme, "What Science Is and What It Is Not," in S. J. Doorman (ed.), *Images of Science: Scientific Practice and the Public* (Aldershot, Netherlands: Gower, 1989).

30. C. P. Snow, *The Two Cultures: And a Second Look (1959)* (New York: Mentor Books, 1963).

31. Gouldner, *op. cit.;* Chap. 6.

32. J. Herbig, *Kettenreaktion. Das Drama der Atomphysiker* (Munich, dtv, 1979), pp. 315ff.

33. Johann Wolfgang von Goethe, *Theory of Colours* (Cambridge, Mass.: MIT Press, 1970).

34. See G. Böhme, "Leib: die Natur, die wir selbst sind," in G. Böhme (ed.), *Natürlich Natur. Über Natur im Zeitalter ihrer technischen Reproduzierbarkeit* (Frankfurt: Suhrkamp 1992).

2

What Makes Us Feel We Need a Theory of Science?

A Note on Terminology

Unfortunately, at the beginning of this essay we must say a word about terminology. I would have much preferred to start with some striking opening sentence such as, "Science is as old as human history" or "Science is becoming the main force behind the violent development of our society." . . . But what do I mean by "science"? Since German is my mother tongue, I cannot but mean "*Wissenschaft,*" when I say "science." *Wissenschaft* includes natural science, the social sciences and the humanities as well. I want to speak about this whole complex comprised by science, the humanities and technology. I have noticed that some English authors have also felt the need to use the term "science" in a very broad sense: e.g., Maslow in his book *Psychology of Science,*[1] is mainly concerned with psychology; or to give another example: Medawar in *Advice to a Young Scientist,*[2] defines science as any explorative activity. It is a remarkable fact that in the German tradition, where there is a common word for this complex domain, there has been such great interest in distinguishing *Geisteswissenschaften* from the natural sciences. Today two facts make it reasonable to begin with a rather comprehensive concept of science: In the first place, the cognitive paradigm of modern natural science has penetrated all other academic knowledge; secondly, as academically trained intellectuals we find ourselves in a historical situation in which we must develop a consciousness of our role in society. But with this remark we are already broaching the main text. We shall, then, make a fresh start.

What sentence should we take as a "great" sentence fit to begin with? As the title of this introduction has already indicated, I do not think it goes without saying that something like a theory of science exists. Therefore, let us put forth the following: *The very existence of theories of science indicates that something is wrong with doing science.*

There is no question that we have a theory of science; in fact, we have several. In addition to the extended investigations in scientific methodology, the theories of the scientific community, the structures of scientific development, the scientific-technical revolution, and others should be mentioned. Our opening thesis throws a strange light on all of these. One may ask whether it is not quite natural (i.e., whether it needs explanation) that we are designing theories of science. We theorize about everything: There is a theory of the world system, of chemical combination, of mental development—but there are also theories of nutrition, nursing, and skiing. Problematizing the existence of a theory, on the one hand, raises the question of what theorizing actually is, and, on the other hand, introduces a psychodynamic perspective. We do not simply accept that because man is a curious and knowledge-breeding animal, he will finally produce a theory of knowledge, or for that matter, a theory about anything at all. My supposition is connected with the fact that theorizing is not just knowing. It is a special type of knowing—incorporating distance, a contemplative attitude, toward the object known, as well as words, arguments, discourse, and a difference between facts and the ground from which they are to be explained.

If this is not satisfying, there is another way of arriving at a critical supposition concerning theories of science: We can begin with the simple fact that there are people—a considerable number of practicing scientists among them—who say that a theory of science is useless![3] It is true, they say, that you have to learn a great deal to become a scientist. But you learn science by practicing science, not by attending lectures on methodology. The transmission of scientific competence is accomplished tacitly.[4] Moreover, a principal problem that should be raised is that of the pragmatic bearing of any theory of science. What the theory says about science must be either descriptive or prescriptive. If what it says about science is descriptive, it must be the result of studies on how science functions—or better, on how it has functioned so far. Hence it is knowledge about a past period as part of a rather innovative, ongoing process, and its value for the future will not be esteemed very highly. If what the theory of science says about science is prescriptive, the question arises as to who the person is who makes the prescriptions. The practicing scientist will scarcely accept prescriptions for his work from philosophers or sociologists of science, and his own colleagues will scarcely form theories about science; rather, they will practice it. Thus, I think there are some good reasons for belittling the relevance of a theory of science for its practice.

The use of a theory of science is obviously not related to the question of how to conduct science but rather to problems of how to talk about science. It responds to the kinds of uneasiness practitioners or the general public feel about science. Let us consider some classical examples—namely, the philosophies of Plato, Kant, and Popper.

Plato must be regarded as the first to develop a theory of science. In fact, he described quite clearly the hypothetical deductive method exhibited by the mathematical sciences of his time; he argued about the relation of scientific truth to sense data; he discussed the use of experiments for the cognition of nature, etc. In doing this he introduced the differences between proof and persuasion, science and belief, the intellectual world and the sensible world, and he thereby irreversibly shaped the history of ideas to come. Why was Plato so concerned with these differences? The answer can be found in the social history of fifth- and fourth-century Athens. When seen within this context, it becomes clear that Plato's demarcations were strategies of exclusion.[5] Plato and his Socratic contemporaries were involved in a struggle about the structure of the newly developing higher education in Athens. Beginning around the time of Pericles, a certain need was felt for education that transcended the traditional basic training in gymnastics, music, grammar, and calculation. The complicated political affairs of the Athenian empire (the Delean League) and the development of democracy made it necessary to be trained in certain competences to fulfill one's political life. In this situation the students of Socrates, the Sophists, and the rhetoreticians were all competing for position within this new policy-related education. Plato's strategy was to distinguish philosophy from mere rhetoric, true knowledge from mere belief, essential reality from mere appearance.

My second example is Kant's philosophy of science. Strangely enough, Kant is not well known as a theorist of science. The reason for this is that he always speaks of "experience in general," although what he in fact does is to give a foundation to Newtonian physics. When considering the Kantian philosophy of science, it becomes necessary to correct one possible misunderstanding in talking about the "uneasiness," "problems," and "trouble" with science as a source of theorizing about science. It is true that in Kant's case as well, a certain uneasiness forms the background of his research. But it seems to be a "positive" uneasiness—namely, the philosophical wonder well known since the time of Aristotle. Kant wonders how natural science can be so successful in establishing laws of nature. In some sense it is exactly the opinion that science can be taken for granted which forms Kant's point of departure—a positive attitude. Only if one looks more closely does the same feeling reveal another component: astonishment about the possibility of science arises from a feeling of being cut off from nature.[6] Since nature is seen as alien to man, the idea that a science of nature is possible needs explanation. Kant's solution to this problem is well known. What we know is not nature as it may be in itself, but as it appears to us. In elaborating this solution, Kant has given very important insights into the subjective side of science, the modes of data generation, concept formation, and theory building.

My third example is Sir Karl Popper's philosophy of science. The emotional background of this theory of science might not be too obvious because it is so close to us. Nowadays Popper's theory seems to be nothing but the true theory of science, which consequently needs no explanation. But there is some indication that the days of Popper's theory as the dominating theory of science are numbered; therefore, we may try to identify its motives. To my mind the irritation at the origin of Popper's theory is the experience of science as having a history; by this I mean a history in the strong sense. The theory of science in our century was motivated by the experience that fundamental principles of science could be decommissioned and that theories—the truth of which seemed unquestionable—were overcome by other theories. It was the emergence of the theory of relativity and quantum theory and the dethronement of Newtonian mechanics which gave rise to the modern theory of science. Popper, in his intellectual autobiography, says: "The decisive point of all this, the hypothetical character of all scientific theories, was to my mind a fairly commonsense consequence of the Einsteinian revolution, which had shown that not even the most successfully tested theory, such as Newton's, should be regarded as more than a hypothesis, an approximation to truth."[7] Considering this background, one can say that Popper turned a distress into a virtue. For him the fallibility of science is its virtue. His theory was the first to consider science as a process; it was published under the slightly misguiding title *The Logic of Scientific Discovery*.[8] Actually, it was not the context of discovery but the context of verification that came under scrutiny with Popper.[9] His contributions belong to a clarification of what constitutes a possible argument within the ongoing debate concerning the improvement of scientific theories.

Looking back at our three examples, we can say that the first revealed the difficulties experienced by Greek intellectuals in defining their place within fifth- and fourth-century Athenean society as a motive for theorizing about science. The second example hinted at the relation of man to nature as a source of irritation that may be the origin of a theory of science. The third example showed the historical self-experience of science that made it necessary to move to the metalevel of research.

It is worth keeping these possibilities in mind as we now turn to our own situation. If we allow ourselves for a moment to reflect upon all our doubts, fears, irritations, bad consciousness, and anxiety about science, it seems to be impossible to continue working in science at all. At least it seems impossible to design a comprehensive theory that supplies an orientation for all the contemporary problems connected with science.

A Theory of Science?

There is the growing overall function of science within our society, which has caused some authors to call science "a" or even "the" dominant force of production, and which has caused others to ask whether intellectuals are on their way to class power; there is the so-called incorporation of science, which describes the fact that science is to a large extent incorporated into economic—and that means for us capitalist—contexts; there is the deeply felt ambivalence toward the relation of scientific progress to the improvement of human living conditions; there is also the obvious destruction of nature by science-based production and traffic; the militarization of science, which means that today between one-quarter and one-half of the investments of human and economic resources in science serves to increase the destructive forces of mankind; furthermore, the scientization of all sections of social activity, which results in preventing people from governing their own lives, making them dependent upon experts; and lastly, there are ethical problems connected with the new possibilities of experimenting with life and manipulating human beings. Thus we are in a situation where science is the source of a great deal of uneasiness, and we are urgently in need of saying what this bewildering human enterprise called "science" is; that is, of formulating a theory of science. This theory must take up the results of Plato, Kant, Popper and others—their questions are still ours. But we can no longer restrict its scope either to the epistemological, methodological, or sociological level. As it has to answer the problems we have with science, it has to be related to the corresponding contexts from which these problems arise. Let us try to outline its different dimensions.

1. A theory of science in our day has to make explicit what science is within a broader spectrum of kinds of knowledge. In fact, this is what most theories of science have done, but they have done so by "demarcations," that is, by defining science as opposed to, say, metaphysics, belief, superstition, etc. But the advocates of these strategies of demarcation were themselves unaware of what they were doing, suppressing as they did awareness of the social impact of their strategies. By establishing hierarchies among types of knowledge, they established hierarchies between the corresponding bearers of the different sorts of knowledge. Moreover, these theories just pushed the other types of knowledge into the background and labeled them irrational, vague, and deceptive. A modern theory of science has to restore to these other forms of knowledge a right of their own; only in this way can it answer one of the fundamental questions connected with science—namely, what the scientization of our lives means.

2. A theory of science in our day has to reflect the relation of science to its object. Strangely enough, we do not have such a theory of science. We have elaborate methodologies and intellectually fascinating theories

of the social system of science, but none of them reflect in any sense that science implies a rather particular relation to its object. Here Kant's epistemology might provide some help, because, as I said before, irritation about the relation of man to nature was one of its motives. But, in addition to the alienation from nature so strongly felt during the eighteenth century, today we have to face the environmental problem and ask ourselves if science itself has contributed to that problem. We can no longer treat science as a merely contemplative activity. We have to ask what science does to its object; we have to conceive science not only as an intellectual but also as a material appropriation of nature.

3. A theory of science in our day has to be at the same time a theory of its subjects, of the people doing science. Indeed, substantial work has been done in this area, namely, in the sociology of the sciences. But what is lacking is precisely the subjective approach—a way to account for the fact that it is we who are doing science. Sociology treats scientists like a population of mice, "observes" what they are doing, and tries to discover the rules of their activity. It does not suffice when knowledge for the sake of orientation is sought. A theory of science that can serve this task has again to take as its point of departure Weber's "science as a vocation"; it has to deal with the tension between science as a vocation and science as labor, and then has to contribute to the intellectuals' search for their position in our society. This part of the theory has to cope with the most difficult problems, not only because it has to overcome the traditional dichotomy between the subject and the object of research, but also because it has to cope with the ideologies through which scientists legitimate their role in society although the ideologies may be mystifying even to them. For example, it is a common ideology that scientists work for other people, i.e., the general public. By this ideology they suppress the consciousness that they have interests of their own, which might even be considered class interests.

4. A theory of science in our day has again to take up the question of truth. I already mentioned that the dominant theory of science today is Karl Popper's. He handles the question of truth merely by suppressing it. It remains one of the strongest motives—perhaps ideological—for doing science, that it is understood to be the search for truth. In fact, science has many other goals, such as usefulness, security of life, domination of nature and society, etc., but one will never understand science so long as one does not understand what the relation of these other goals is to truth. The honorable self-effacement that science displays before the question of truth does not help as long as it has not been stated what truth is and what type of knowledge might be better equipped than science to reach truth.

5. A theory of science in our day obviously has to produce an understanding of its history and the process of scientific change. This question had already motivated Popper's theory, but finally came to the surface with Thomas Kuhn's book *The Structure of Scientific Revolutions*.[10] Here one can already find discussions of the concept of progress itself. The concept of progress originated during the Renaissance,[11] and scientific development was actually the main example that gave meaning to this concept. Not only has the hope expressed by early ideologists of science such as Francis Bacon that human progress would be connected to scientific progress been questioned today, but so has the very concept of scientific progress itself; Kuhn considers the possibility that it might not be progress in the full sense, that is progress toward an end, but rather only progress away from an origin. In addition, there have been some frightening signs that the enterprise of science might not be a march to an endless frontier[12] but a task for mankind which might one day come to an end.

6. A theory of science in our day cannot cope with science as a separate system within society. It is true that any investigation of some particular subject tends to isolate it from its environment. But in this case we would lose insight into one of the most upsetting questions concerning science—namely, its overall function within our society, i.e., the transformation of our society into a technological or knowledge-oriented society—where knowledge means in essence scientific knowledge. A theory of science actually has to be a theory of advanced societies as a whole. We already have some good forerunners in this direction, e.g., Daniel Bell's theory of the post-industrial society[13] and a theory of the scientific-technical revolution.[14] But these examples show that a theory of society as a whole, on the other hand, is not a theory of science. In consequence, a theory of science must be conceived in such a way that it can function as part of a larger theory of society.

It should have become clear that a theory of science conceived in this fashion will mainly function as an attempt at self-understanding by people who are in this field. It should answer to the manifold uneasiness that we feel about science today, a science that we are nevertheless attached to or enamored with. Thus, it should be a kind of knowledge that provides orientation and enhances self-consciousness and self-understanding.

Before going into particular topics related to our concern about science, it may be appropriate to give a preliminary characterization of science. This should not be understood as a definition of science. A definition of science that fits our purpose—conceiving of science in a way that enables us to cope with its reality in our societies—can only be given through the development of a theory of science. But it would aid in our understanding of what follows to give at least an introduction to what I intend by "science."

Science is a collective production of knowledge. This production is performed intentionally and in accordance with rules. Its fundamental methodology consists in the dual methods of analysis and synthesis. It therefore stands on two legs: data production (analysis) and concept or theory formation (synthesis). Thus, as a way of knowing it is characterized by reconstruction: Something is known when it has been reconstructed on the basis of data organized by certain synthetic procedures.

As a type of knowledge, science is impersonal, i.e., detached from the person doing science. The individual is not the vehicle of science but, rather, a contributor to or a participant in science. The results of science are objectified; that is, they are stored by structuring material things, stamping things symbolically (books, etc.) or concretely (machines, etc.). Therefore the tendency in science is toward a reconstruction of the world.

Notes

This chapter first appeared in the *Graduate Faculty Philosophy Journal* 12 (1987): 3–11. © 1987. Reprinted by permission.

[1.] A. H. Maslow, *The Psychology of Science* (New York: Harper and Row, 1966).

[2.] P. B. Medawar, *Advice to a Young Scientist* (New York: Harper and Row, 1981).

[3.] See G. Holton, "Do Scientists Need a Philosophy?" *Times Literary Supplement,* November 2, 1984.

[4.] See M. Polanyi, *Personal Knowledge* (New York: Harper Torch Books, 1962).

[5.] For further explanations, see my paper "Demarcation as a Strategy of Exclusion: Plato and the Sophists," in *The Knowledge Society,* ed. G. Böhme and N. Stehr, *Yearbook of Sociology of the Sciences,* vol. X (Dordrecht: Reidel Publishing Co., 1986).

[6.] See my paper "Kant's Epistemology as a Theory of Alienated Knowledge," in *Kant's Philosophy of Physical Science,* ed. R. E. Butts (Dordrecht: Reidel Publishing Co., 1986).

[7.] "The Autobiography of Karl Popper," in *The Philosophy of Karl Popper,* ed. P. A. Schilpp (La Salle, Illinois: Open Court, 1974), p. 64. There is still another source of Popper's philosophy—namely, his relation to Marxism. The shooting of some young socialists, who according to Popper had followed a suicidal policy in believing that Marxism was "scientific," was an existential experience for Popper in his early period in Vienna. See the above autobiography, pp. 25–28.

[8.] Karl Popper, *The Logic of Scientific Discovery* (London: Hutchinson, 1959).

[9.] This distinction has been introduced by H. Reichenbach.

[10.] Thomas Kuhn, *The Structure of Scientific Revolutions* (1962), second ed., eds. O. Neurath, R. Carnap and C. Morris, *International Encyclopedia of Unified Science,* vol. II, 2 (London: University of Chicago Press, 1970).

[11.] See W. Krohn's contribution in G. Böhme, W.v.d. Daele, and W. Krohn, *Experimentelle Philosophie* (Frankfurt: Suhrkamp, 1977).

[12.] See V. Bush, *Science: The Endless Frontier* (Washington, D.C.: U.S. Government Printing Office, 1945).

[13.] D. Bell, *The Coming of Post-Industrial Society: A Venture in Social Forecasting* (New York: Basic Books, 1973).

[14.] R. Richta et al., *Civilization at the Cross-Roads,* third ed. (White Plains, N.Y.: International Arts and Sciences Pv., 1969).

3
The Formation of the Scientific Object

It may not be obvious to everyone that it should be possible to characterize science by its object. The common sense opinion of science is that it is knowledge about any object, and that its aim is simply to say what is. This general opinion implies two suppositions: namely, (1) that the realm of science is unrestricted, comprehending the totality of being, and (2) that the relation of science to its object is merely a receptive one. Both suppositions are wrong.

One way to approach the truth about science is to notice that science is not about everything; for example, there is no science of the individual. It is very often the case that science is made to handle certain individuals, for instance, the ecology of New York; but what is scientifically known about the individual is known in terms of universals, that is concepts, and hence pertains only to certain aspects of the entity concerned. Classical philosophy used to formulate it this way: *individuum est ineffabile*—you cannot tell what the *individuum* is as such. This formulation seems somewhat exaggerated, at least if by "tell" one means any possible use of language whatsoever. You can address an individual entity in words, and as everybody knows we talk *to* individual persons. However, one cannot leave the realm of universality by talking *about* something.

Science is not about the *individuum*. This has led some authors to postulate another science (e.g., Maslow, op. cit.),[1] or to put forth strong criticism of scientific rationality. Thus Horkheimer and Adorno in *Dialectics of Reason* blamed scientific rationality for leading to the suppression of the individual in our century.[2] But this criticism only mirrors the fact that science was taken to be more than it is and was attributed more functions than it really has.

Science is not about the *individuum,* nor is it about the whole—the totality of being. This must be remembered, if we are not to expect a world view from science. Since the Renaissance, science has functioned with respect to world views, but this function was a critical one. There is no science of the whole world; the object of science is (in general) always something isolated from the rest of the world. One objection to this characterization of science is that there is a physical cosmology. But this argument only indicates that the process of isolation in forming a scientific object should not be understood too narrowly, i.e., in a spatial sense; in fact, the objects of cosmology are rather abstract models of space, e.g., matter-densities, etc. But in no sense is science about the totality of being.

If science is not about the individual, nor about the whole world, is science then about things? Or events? The answer, again, must be "no." It is true that the terms "thing" and "event" can be understood in such a way that one can say that science is about a certain kind of thing, or a certain kind of event; e.g., one can say that two atoms of H and one atom of O have combined to form one molecule called "water." But these are not the kind of things we are used to calling "things" in everyday life. The expression "water" makes this clear: The chemical formula mentioned above treats water as a molecule. But water as a molecule is quite a different thing from what we know water to be in everyday life. For example, it is not liquid. It is a highly complicated task to reconstruct scientifically a phenomenon like liquidity on the grounds of the interaction of a very large number of molecules. It is again a common sense notion that science is about the same thing as everyday knowledge—in the same way, but better and more exact. It is one of the purposes of this essay to demonstrate that science diverges from everyday knowledge.

The last negative point I want to mention is that science is not about nature. This may be the most surprising statement, for wasn't the definition of science meant to be knowledge about nature? Indeed, the formulation is paradoxical, and deliberately so. For it should be clear that the object of science is neither "nature out there" nor nature in a sense where we ourselves are nature. Clearly science is in some sense about nature, but nature as the object of science is a highly sophisticated entity—the result of a process, the making of the scientific object. Now let us follow this process in historical and epistemological detail.

The emergence of modern natural science is usually described merely as a progressive event: that Aristotelian nonsense was put aside, scholastic boredom overcome, and medieval superstition refuted. It was not until recently that attempts were made to evaluate Aristotelian natural science on its own terms,[3] and that the question of what man's relation to nature was before the scientific revolution was raised.[4] To be sure, the emergence of modern natural science was a process of enlightenment. But where new light appears, there is also new shadow—an insight formu-

The Formation of the Scientific Object   31

lated by one of the great figures of the Enlightenment, D'Alembert.[5] It is rather unlikely that the ideas of great philosophers and scientists before the scientific revolution were just nonsense. It is time to raise the question of what has been lost through the scientific revolution, what types of knowledge about nature came out of practice, and what sort of insights into nature has been lost. Only when these questions are answered at the same time can the formation of the object of science as a historical process be understood in its specificity.

I cannot draw a picture here of this historical process as such; I shall instead give an idea of the main changes in our relation to nature, ordering the particular elements into three clusters.

The first cluster concerns man's self-understanding with respect to nature. One can summarize the changes by saying that they imply the loss of the human body as a paradigm for knowledge. Even we members of the modern world willingly agree that we belong to nature—that we are ourselves "natural"—but we do that without drawing any epistemological consequences. To give an example, we do not think that we know something about the world from what we have experienced with or about our own bodies. This was fundamentally different before the "great instauration"[6]—the microcosm-macrocosm analogy was one of the most important epistemological principles. This principle, which seems rather exotic to us (e.g., when we are confronted without explanation by one of the pictures showing the cosmos as Great Man) contains, as far as I can see, two moments. First, the cosmos is seen as alive—that is, as an organic, self-moving entity. This idea goes back to Plato's *Timaeus*. Secondly, sympathy (*sympatheia*) is estimated to be one of the cognitive faculties of man. In ancient times people made philosophical currency of the fact that in our own bodies we can feel what occurs in nature around us or in a fellow natural being—to a certain extent at least.

The second cluster is again concerned with the epistemological position of the human body. But here we are not concerned with the use of self-experience for knowledge about nature but with the gradual exclusion of personal experience from relevancy for "objective" knowledge at all. It is true that viewed historically this is a long process. But this tendency is one of the most important characteristics of modern natural science. On a very general level, one can say that for natural science personal experience became more and more irrelevant; whereas in former times, "experience" meant personal maturity, that one was well-travelled and had accumulated a great deal of knowledge. This biographical embedding of knowledge had to disappear in order for experience to be valuable within the new experimental philosophy:[7] Experience now came to be understood as that which could be perceived by anyone following certain rules of experimentation.

This abstraction from the personal context meant that natural science strove as far as possible to rid itself of the human senses as a medium of experience. In its essence, modern natural science is not based on sense-perception but on measurement—that is, instrumental experience. This process of replacing the human senses by instruments is a very characteristic one. Most cases of early modern science begin with instruments that support or extend the human senses. But after a certain period, there is a turn toward instruments to *define* the phenomenon or effect to be observed, so that the question then arises whether the human senses are appropriate to measure or estimate the phenomenon or effect under consideration. The development of temperature measurements, e.g., the development of the thermometer, is a case in point.

This predominance of instruments within scientific experience had far-reaching consequences for the behavior of scientists and for the understanding of the human sense organs themselves. Scientists had to adapt themselves to instruments, to accommodate their sense-perception to the functions of instruments; thus the human sense organs themselves were finally understood on the model of instruments.[8]

The third cluster of changes in the relation of man to nature is most distinctly mirrored in the rise of physico-theology in the eighteenth century. Physico-theology was in some sense the product of the thinking of the experimental philosophers themselves—Newton and Boyle in particular. They left to the theologians what they could not explain within the framework of modern natural science. Thus the Boyle Lectures became the home and the origin of a type of theology that contained traits of nature which no longer had a place within the science of nature and which were thought to prove the existence and the excellence of a divine creator. These traits were the wholeness, the quality, the meaningfulness, and the beauty of nature. All these characteristics of nature had no place within a natural science which tried to explain nature as an aggregate of pure facts. But they were methodological guidelines or comprehensive expressions of the science of nature in earlier times. Consequently, physico-theology in the eighteenth century served as a treasure house of human knowledge about nature that could not be integrated into the emerging new science: knowledge about purposeful relations within or between organisms, knowledge about animal behavior, knowledge about circles of reproduction within nature and other systemic relations, i.e., knowledge about nature that today we are again trying to develop under the title of ecology.

We can summarize what we have achieved thus far as follows: Modern science does not proceed on the basis of any similarity between the subject and the object of knowledge, i.e., on the basis of any similarity between man and nature. Modern natural science is "insensitive"; it is unaffected and instrumental experience. Modern science is science about mere facts, which excludes the unity, beauty, quality, and meaning of its objects from its range of consideration.

One can also summarize these facets of the relation of man to nature by saying that within natural science nature is conceived as alien to man. This formulation calls for a historical explanation: How did nature become alien to man, where does the gap between man and nature which is constitutive of modern natural science come from?

These questions open up a new approach to the development of modern science. The received explanation claims that modern science descends from a marriage of traditional (scholastic) scholarship and artisanship. This is a good explanation; it refers to the social and historical background of the scientific revolution and accounts for the affinity between modern science and capitalism. It can also explain the pre-established harmony between science and technology which is so obvious in our day. The very decisive turn made by Galileo was to overcome the ancient dichotomy between nature and technics. For Aristotle and all occidental philosophers following him, technical achievements were something *para physin,* i.e., something lying beside or even going against what happens naturally, namely by itself. Galileo, one of those Renaissance figures who combined scholarship and technical skills, realized that what mechanics performs resides *within* the framework of nature, and is achieved in accordance with nature. His approach to understanding mechanics from within the perspective of natural science on the other hand led him to understand nature from a mechanical point of view. So much for the traditional explanation.

But another factor has to be added today, one which accounts for the characteristic alienation of nature in modern science. Who were the practitioners of science in its early days? In answering this question one notices that in considering the social and historical background of modern science, one branch of development has been neglected: Perhaps under the influence of Marxist historians, the social origin of modern science has been identified with the rising bourgeois class. This element was certainly there. But another important origin is to be found within the royal courts. An important part of early modern astronomy had its place in these courts, and the same is true of instrument making, biological classification, and experimentation in many other fields. From the early sixteenth century onward it was fashionable at the courts to have a curiosities cabinet, a physical cabinet, a bestiary, or a botanical garden, and a court astronomer employed as astrologist. The custodians of those cabinets, the directors of the royal gardens (e.g., the *Jardin des Plantes* in Paris), contributed greatly to the knowledge of nature, the courts' mechanics to the invention of scientific apparatus, instruments, and the discovery of certain effects. For the artisans at the courts were not bound to the rigid rules of the guilds. On the contrary, they were continuously encouraged to produce "effects" and "phenomena" for the curiosity of the feudal class.

Having supplemented the well-known view, we can now say what was characteristic for those doing science in early modern times: They had no immediate contact with "nature out there," with nature as it is in itself, and they also did not rely on the experience of those who had such contact. Even the crafts, which according to the received view were constitutive of modern science, were of a particular kind. It was not the skill of farmers, sailors, miners, or hunters that played a role in science; it was rather the skill of the glass-grinders, watchmakers, engravers, gunmakers, etc., that played an important role in this process. Nature—as far as it passed through their hands—was already "appropriated" nature, i.e., isolated, separated, purified, specified pieces of nature. Their relation to nature, and consequently the relation of modern scientists to nature, was also a mediated one—mediated through the work of those who struggled in immediate contact with nature. Thus we can describe the formation of the scientific object in a very concrete sense. The object of modern science is "indoor nature," nature under well-defined and controlled conditions, nature prepared by human work. What is generally known, namely that natural science is the science of the laboratory, appears in a new light: It means that this kind of science deals with nature under the condition of a prior material appropriation of nature.

The alienation of nature which lies at the origin of modern science means that there is a gap between the practitioners of science and immediately experienced nature. They keep nature at a distance, and their research begins when the struggle with nature is already finished. They are concerned with nature under human conditions, that is, social conditions.[9]

As a consequence, we must understand natural science epistemologically as alienated knowledge—that is, knowledge of what is essentially alien to the knower. The epistemological problem arising here is well known from ethnology: How can you understand a culture that is fundamentally unfamiliar to you? This ethnological problem is not only comparable to but also contemporary with the development of natural science in the narrower sense. For the discovery of new worlds and the confrontations of Europeans with strange cultures belong to the same centuries. The solution to the problem was the same—namely, one understands a stranger by "appropriation," by observing him under conditions one has set, and by bringing him into one's own framework. Let us consider how this works on the cognitive level.

Before making a fresh start it may be worthwhile mediating for a moment on what a peculiar and astonishing thing it is to acquire knowledge about something that is absolutely unfamiliar to oneself. What does "to know" mean in this case? How do we bridge the gap so as to get in contact with the object of knowledge?

For us it is certainly evident that there is a primordial connection between man and nature, in virtue of which man is a part of nature. The problem at stake is a real problem only for those who contest the natural status of man, or at least the epistemological functions of his natural components. This was true for the philosophers of the seventeenth and eighteenth centuries; thus they were inclined to explain the success of natural science by another similarity between man and nature. Unable to acknowledge that man is a natural offspring of nature, they postulated that somehow there must be reason within nature; thus natural science was the discovery of reason within nature. The positions implied within the idealistic account of the success of science ranged from the deistic belief that nature *is* God, through the belief that nature was created by a divine intelligence whose thoughts it reveals, to the position that the thesis of "reason within nature" is just a methodological hypothesis.

The origin of idealistic thinking resides in the cleavage between man and nature. As modern natural science implies this cleavage, it is itself inevitably idealistic. Although this is the case, it is impossible to give a satisfactory account of the success of natural science on mere idealistic grounds. In fact, natural science would be impossible without the concrete, bodily connection of man with nature. This connection is on the one hand given by his phylogenesis from nature,[10] which means that there has been a primordial adaptation of man to nature; on the other hand, it is given by work and working experience. Although this fundament is neglected in science, it is nevertheless operative and harbors knowledge of nature—concerning what is and what is not possible in nature—before the very scientific research begins.

It is said that science begins with data. In Latin, "data" literally means: the given. But natural science does not accept anything that is given to us from nature just as it is. It is not a flower, or a stone, or water as we find it in nature, that science takes as a beginning. "Data" in a scientific sense means results from measurement or scientific observation—that is, by means of analysis. Scientific data are in no case that which is immediately given, but what is constructed from it by certain methods of analysis. Much has been said about the so-called theory-laden character of scientific data, namely, that certain data are to be only obtained by presupposing a theory—but this is not what we are discussing. Our point is even more fundamental. It is that scientific data are dependent upon certain methods of data generation. Kant called these methods or rules of data generation *Anschauungsformen,* or forms of perception, and with this expression he in fact exaggerated the point. For the clue to these methods is that we do not just perceive nature, we do not expose ourselves to nature or experience nature, we *make* experiences about nature. Maintaining a distance and accepting only what corresponds to certain rules are quite characteristic of natural science. But Kant was right in claiming that, as a consequence of the methods of data generation, we know something *a priori* about nature. Our procedures of data generation impose some general structures on the sets of data to be obtained. These structures can be known in advance, that is, before producing any particular set of data. For example, when approaching nature with a balance you know that the set of data to be obtained will form an additive group. This is the consequence of the functioning of a balance, or better, of the rules in accordance with which balances are handled.

Here we are facing a first instance of *a priori* knowledge about nature, at a point where it is well to remember the concrete fundament of science, which lies in the practice of work. That something like a pair of scales is possible in nature, and that it can be handled effectively, is an experience that comes *before* natural science and is one of its preconditions.

We shall summarize this first step of our epistemological argument: Scientific experience is a highly selective, rule-guided perception. It keeps nature at a distance; nothing happens to the person who gains the experience. It can be called a step in the formation of the scientific object because it is a process of data production which imposes a certain structure on the data. In fact, it is nothing more than a step. For we have so far only a set of data—many scientists will say out of frustration with experience: "a mass of data." This is why Kant says in his epistemology that data as such are not connected, that any combination must be introduced by reason. Let us consider this situation in more detail. Did we not say that the set of data has a certain structure, *a priori*? That is true. We are already in possession of what can be called variables, i.e., different procedures for performing measurements and corresponding structures informing sets of possible data. But we do not have "objects." Furthermore, we know on the basis of everyday or work experience that certain groups of data belong to one and the same "thing," namely, an entity given as one in practical experience. For example, in psychology we may know that sets of data produced by different tests pertain to the same person. But this knowledge may only serve as a stimulus or, at best, as a legitimation to look for some connection between the sets of data— this connection is not *given* by the sets of data themselves, that is, by what is acknowledged as the basis of scientific reasoning. The unity of the "thing" has been lost in the analytical process of data production. This unity must be reestablished by scientific reasoning as the unity of the scientific object. To give some examples: Two sets of psychological data must be ordered in accordance with the scheme of motives and performances—on the basis of the idea of a person acting in conformity with certain motives; or two sets of data obtained from physical measurements, for instance, positions and velocities, must be ordered on the basis of the idea of a mass having changing states of movement. We can call these ideas "concepts of objects"—they form part of what is known as scientific theories. Concepts of scientific objects, as a rule, contain not only a certain combination of variables, stating which sets of possible data characterize the object, but also the exclusion of other variables, which are thus seen not as representing the object but as representing the rest of the world over against it. Thus we have a formation of the scientific object from the conceptual or theoretical side, what Kant called the constitution of the object by pure concepts of reason. They are responsible for drawing the distinction between object and world, and constitute the unity of the object. Further analysis following Kantian epistemological lines would reveal that they are also responsible for the way we conceive of the identity of the object through time.[11] Another very important function of these concepts consists in their giving the general structure of laws of nature. As this has been very often misunderstood—in particular because Kant gave his relevant statement the provocative turn encapsulated by "we prescribe the laws to nature"—we should give some more attention to this point.

In some sense it is obvious that what we want to learn from nature is just the laws of nature—this is one of the purposes of doing science. It is also obvious that nature by herself reveals regularities. The manner of conceiving of these regularities is not given by nature, but rather the pattern of natural laws is contributed by the knowing subject. To give an example, since Newton mechanical laws have been conceived as exhibiting the characteristic pattern of a relation between a force and a change in the mechanical state of the object. This concept of a law presupposes on the one hand what a mechanical state is and what it means to remain in the same state, and on the other hand that a force is conceptualized as the cause of a change of state. These two presuppositions are laid down by Newton's first law, that a body remains in its state of rest or uniform rectilinear movement so long as this state is not altered by external forces. The idea of inertia is constitutive of what a mechanical state is and consequently of what an external force is. Newton's second law provides a rule for calculating forces from alterations of state and vice versa.

Taking together the empirical and the theoretical sides, we can say that the formation of the scientific object is a process of reconstruction. What has been given by everyday life experience and the practice of work is analyzed on the empirical level and differentiated into a certain number of variables, or parameters. The unity of the originally given thing is thus lost. It is reconstructed as the unity of the scientific object. This is done with concepts, which allow one to arrange the variables in a certain order. The behavior of objects is then accounted for by laws, the form of which consists of a pattern enabling one to conceive of the regularities which the data might reveal.

In this summary we have stressed the *a priori* moments of the processes of knowledge production. But it is true that the *a priori* concepts governing the reconstruction are not without empirical grounds. They cannot be formed arbitrarily. But the presupposed experience is not the scientific experience. For before starting with scientific experience, one must already have prepared one's instruments, one must have decided upon certain measuring procedures, and one must have differentiated among variables. On the other hand, before accounting for one's knowledge about regularities in nature, one must have decided where the dividing line between one's object and the world is to be drawn, that is, which data characterize the object and which do not; one has to decide what a state is and what change is, and one has to pattern the general structure of scientific laws.

This epistemological picture of the formation of the scientific object fits quite well with the results yielded by the social and historical approach. The differentiation among variables corresponds to the development of instruments; and the differentiation between the object and the world corresponds to the isolation of pieces of nature in the laboratory and the control over border conditions. The material appropriation of nature is reflected by an intellectual appropriation: Nature is conceived within and under human conditions.

One question addressed in this chapter pertained to what the object of science is and how science is characterized by its object. We can now say that science is a kind of knowledge which keeps its object at a distance from the subject of knowledge. The latter should in no way be affected by its object. This is accomplished by very rigid procedures of scientific experience and reasoning. By these procedures the object is submitted to the conditions of the subject of knowledge; they form the design for a construction of the object. As this construction is not mere fancy but has constraints in the material, one may say that scientific knowledge, by its essence, is reconstruction. Reconstructive knowledge is very effective and exact knowledge. This is the advantage derived from a clear definition of the border conditions. But a reconstruction is not the original thing. Science is not about things. As we have said before, science is not about nature; rather, it uses nature as the material for a step-by-step reconstruction of the world. This means, however, that the world is reconstructed step by step on a practical or technical level.

Notes

This chapter first appeared in the *Graduate Faculty Philosophy Journal* 12 (1987): 13–24. © 1987. Reprinted by permission.

[1.] P. B. Maslow, *The Psychology of Science* (New York: Harper & Row, 1966).

[2.] Max Horkheimer and Theodor W. Adorno, *Dialectic of Enlightenment* (New York: Social Studies Association, 1944).

[3.] See "Aristoteies' Chemie: eine Stoffwechselchemie," in G. Böhme, *Alternativen der Wissenschaft* (Frankfurt: Suhrkamp, 1980).

[4.] Carolyn Merchant, *The Death of Nature* (San Francisco: Harper and Row, 1980).

[5.] *Eléments de Philosophie I, Mélanges de Littérature, d'Histoire et de Philosophie* (Amsterdam, 1758), vol. IV, p. 1.

[6.] See Francis Bacon's *Instauratio magna* (1620).

[7.] G. Böhme, W.v.d. Daele, and W. Krohn, *Experimentelle Philosophie* (Frankfurt: Suhrkamp, 1977).

[8.] For this cluster, see the doctoral thesis of W. Kutschmann, *Der Naturwissenschaftler und sein Körper,* Darmstadt (Frankfurt: Suhrkamp, 1987).

[9.] For an elaboration of this view, see Hartmut Böhme and Gernot Böhme, *Das Andere der Vernunft* (Frankfurt: Suhrkamp, 1983), chapters I and V.

[10.] This is the proper field of an "evolutionary epistemology."

[11.] G. Böhme, "Towards a Reconstruction of Kant's Epistemology and Theory of Science," *The Philosophical Forum,* vol. XIII, 1981, pp. 75–102.

4

Can Science Reach Truth?

The formulation of the title above is not the most appropriate one if the question refers to the search for truth. "Is science true?" or "are scientific theories true?" or "is science true knowledge?" would to my mind be much better expressions for what we really want to know. The question framed by the title mirrors Popper's philosophy or rather his doctrine that all scientific knowledge is merely hypothetical. It is an honorable virtue always to keep in mind that to err is human—but does this preclude that human beings are in possession of some truth? In fact, Popper does not eliminate truth from the domain of science as do some of his followers, such as Larry Laudan: "Science does not, so far as we know, produce theories which are true or even highly probable."[1] For Popper, truth remains the ultimate goal of science. Science, according to him, is "conjectural knowledge," and its work is to improve conjectures in order to make them more "truth-like." Therefore, on Popper's view, the search for truth becomes a question of approximation *to* truth.

This is the point at which our own question takes its departure: The concepts of conjecture and truth-likeness are meaningless, if one does not have an idea of what it would mean finally to reach truth or even merely to hit upon it—if one does not have at least one criterion to detect whether what one has is true. In my opinion Popper's philosophy is not only lacking such an idea but even excludes it; for the method of trial and error, of which Popper's scientific methodology is a sophistication, is not appropriate to reach truth at all. More generally speaking, any evolutionary epistemology, which conceives of knowing as a mode of adaptation, has nothing to do with truth.

I concede that this statement presupposes a particular concept of truth, but I would claim that it is a concept which does not skeptically reduce its requirements at the very beginning. As a preliminary formulation, we shall state that truth amounts to saying what things are and how they behave. This means that the idea of truth implies a certain transparency on the side of the object of knowledge, and insight or understanding on the side of the subject of knowledge.[2] Thus adaptation is not the right way to reach truth, for adaptation is essentially blind. Through adaptation one is only related to surfaces, to effects; one can at best produce a negative of what is. But one is bereft of the possibility of converting it into a positive because one does not know the code for transforming the negative into a positive. Adaptation is like groping through a curtain which is never lifted. However, there are people who say that this is exactly the fate of science. But is it really?

Due to Popper's philosophy, which has nearly become the ubiquitous ideology of science, many scientists would even say that there is no truth in science, that there are not even theories, but that there are merely models. On the other hand, they take a great deal of knowledge for granted and behave during their workday as if convinced that they have a stock of real knowledge at their disposal. However, what scientists believe and what they think about science are empirical questions. As a personal motive for proceeding, I shall mention that when I became a scientist, I was deeply impressed by the fact that laws of nature exist. One experiences even greater astonishment about this if one feels that there is a certain distance or gap between man and nature, that we experience a certain alienation from nature.[3] Paradoxically, as I would like to show later, this distance from nature is the reason why true knowledge about nature is possible.

I must concede, however, that much in natural science is merely hypothetical, and that many theories are just models. But that these aspects of science are the characteristic ones is a perspective produced by philosophers of science, who quite naturally were preoccupied with the esoteric frontiers of science. In our century, philosophy of science is motivated by very far-reaching revolutions which took place at these frontiers, and thus conceives of science as a process of permanent revolution. Even where some quiet times of puzzle-solving are allowed between paradigm switches, as is the case in Kuhn's theory, the idea is that through revolutions the unquestioned basis of these peaceful times is questioned and finally overcome. But when looking at the history of science we might be surprised by the opposite, namely, by the fact that even through revolutions much is preserved. What carries even more weight is a trivial fact that escaped the attention of philosophers, namely, that those overcome and "refuted" theories remain the fundament of a very broad—perhaps the broadest—proportion of the scientific enterprise, i.e., of research directed toward technology and development. Thus classical mechanics, thermodynamics, and electrodynamics maintain a position that is unaffected by the scientific progress that followed their establishment. This gives us the key to a reformulation of our question: for it is the phenomenon of scientific theories becoming *classical* which gives rise to the question whether science might occasionally arrive at truth. Is it possible that scientific theories might be true?

This formulation makes some serious considerations necessary, such as what is required if one wants to talk about the truth of scientific statements (or of theories conceived of as complexes of statements).[4]

The first issue here is that a correspondence concept of truth is indispensable. If anywhere, the consensus theory of truth might have a chance within the social sciences. For in social reality what is, is always a result of consensus. But in natural science the phenomenon needing explanation is just the well-established relation of science to facts; one might say it is the success of science, the technological demonstration that science has really got hold of the objects it studies.

However, in a sense a correspondence theory of truth cannot be sufficient, and this might be the deeper reason why the idea of truth faded away from the philosophy of science. For, on the one hand, in many cases theories cannot be applied to many facts they are supposed to cover,[5] and on the other hand, most of the ones which they are supposed to cover are not facts at all because they belong to the future. One may suspect that here we have once again tapped into the problem of induction. But I shall not give our question the form of how a correspondence between certain propositions and facts might be extended to non-facts—namely, future events. Our problem is: What is the basis of construction? For what science really does is to produce an understanding of nature on the basis of which construction, i.e., technology, is possible. It is an inadequate description of science to say that it gathers experiences which it projects into the future.

Hence, to the traditional achievements of science, according to which scientific knowledge corresponds to the facts and allows forecasting, we have to add a third one, according to which it makes construction possible.

This means that some correspondence must be included in the concept of scientific truth but that this is not enough. The concept of validity must be added: Science makes construction possible because it contains the laws which govern the behavior of certain objects. But what is meant by "validity"? The term "validity" is not unfamiliar in discussions about the value of scientific theories. Therefore, we must be cautious not to connect this term with a meaning unsuited to our purpose. The term "valid" here will not mean being accepted in the scientific (or any other) community. Avoiding this misunderstanding is very important because at this point the concept of scientific truth attracts some elements of the consensus theory of truth[6] without discarding the correspondence point of view. As a consequence of what we have said so far, we must say that since the concept of scientific truth contains elements of correspondence *and* validity, an exhibition of the truth of a scientific statement must comprise a demonstration of its correspondence to the facts *and* arguments for its validity.

But, again, what is meant by "validity"? Like the term "law," its meaning has its origin in the sphere of social life, in particular the sphere of jurisprudence. When I said that the sense of "generally accepted" should be avoided, I meant the social connotation of the phrase. But should it be entirely stripped of this metaphoric sense? It is true that the idea of God informing nature, which in its turn "obeys" these laws, does not help much in this context. But we have to understand that scientific truth consists in certain statements being valid *for a field of objects*. The concept "valid for a certain field of objects" should mean: Scientific statements are valid for a certain field of objects if they formulate the rules which govern the behavior of those objects.

The advantage of retaining the traditional metaphors of law, validity, and governing is that it allows one to bring forward grounds other than empirical ones for the truth of a scientific statement (i.e., measurement or experiment)—namely, *arguments* for why the objects in question must behave as the statement asserts. It may be somewhat puzzling to see the terms "statement" and "assert" used together with "law" and "govern"; as we said before, correspondence is a certain moment within the concept of scientific truth. But here we have to emphasize that the main function of scientific theories is not to make assertions about facts but to formulate the laws or rules governing the behavior of certain objects.

Having distinguished the moments of correspondence and validity within the concept of scientific truth, one will notice that scientists are generally much more concerned with the demonstration of correspondence to the facts than they are with looking for arguments for validity[7]—and so were philosophers of science. Strangely enough, scientists simply trust the validity of scientific laws. Do scientists intuitively know more? Again, what is it that makes scientific theories true? What are possible arguments for their validity?

It would be both unreasonable and unfair not to mention that there is one philosopher who cared about arguments for the validity of scientific laws—namely, Immanuel Kant. Let us return, then, to his epistemology to find out what can be learned from it with respect to our question. One of Kant's main doctrines is that science is not about nature, i.e., about how it may be in itself (*das Ding an sich*), but about nature as appearance. Consequently, truth for him is not correspondence between things and ideas of them, but between representations—namely, between intuition and understanding or between the intuitively represented and the conceptually represented object. The superiority of Kant's philosophy over most recent discussions about correspondence is that it avoids the pedestrian problem of talking about a correspondence between things and verbal expressions, between facts and statements. However, in order to make modern use of the Kantian structure, one must change one side of his correspondence: The empirical representation of the scientific object is not an intuitive one and is not given through the senses but, rather, is a representation of the object by data produced through measurement.[8] In this way, we introduce the technical side of science. However, I do not want to follow this line of thought now.

More important is the Kantian doctrine that the validity of natural laws cannot have empirical grounds, for those grounds could not be compelling. The grounds for the "lawlikeness" of natural laws, or for their character of lawfulness, must be *a priori*. According to Kant, the grounds must establish a necessary relation between the conceptual and all possible intuitive representations of the object. Kant's solution to this problem reads as follows: Understanding determines the perception of the object. Thus, the object is already perceived in such a way that it may subsequently be thought in accordance with the concepts of reason. Again, this idea must be emancipated from psychological metaphors, and the technological aspects of science must be introduced. However, what we have to learn from Kant is that arguments about the validity of natural laws must be concerned with a necessary relation between the empirical and the conceptual representation of the object.

Our next point will be to ask whether such a relation can be found in science. To answer this question I shall give an example—the only possible way it can be answered at the moment. For scientific theories have surprisingly different structures, and the required relation will most probably have a different form for each.

I deliberately take an example from the set of theories labeled "classical." The actual attitude of scientists toward these theories and toward their function as a basis for technology gives rise to the supposition that they might be true. In any case, their historical stability calls for some explanation. My case is that of classical hydrodynamics.[9] In a historical perspective, classical hydrodynamics reveals an additional trait which makes it a favorite object of analysis in our context; as a theory, classical hydrodynamics consists of five equations, i.e., the principle of continuity, the energy principle, and the three Navier-Stokes equations. It is a theory of fluid mechanics. The interesting point is that the theory, which in this form has been complete since about 1850, could not be applied to virtually any interesting cases in fluid mechanics for more than fifty years. However, the scientists working in this field did *not* try to alter the theory. On the contrary, during that period of time, they remained committed to finding ways to apply the theory, that is, to find means of integrating it for particular cases. The long period of inapplicability is very important, because it is a period in which neither verification nor falsification could take place. Thus, the scientists' confidence must have had its ground in something else. They were convinced that classical hydrodynamics was valid for the fluids that interested them, even though they could not apply it.

It might be an idle question to ask what the basis of the confidence in the theory was—perhaps it was just an intuition that it was a "good" theory. But with the benefit of hindsight, we can say that the scientists could have given reasons. Paradoxically, we can say this because it has turned out that their confidence was *not* justified: Classical hydrodynamics is *not* valid for the vague field of possible applications they had in mind. In a very vague sense, this field consisted of "all" fluids. During that period, scientists working in fluid mechanics thought that in principle all fluids were "like" water and air, the most important ones in practical life. Today, we know that not all fluids behave like water and air, but, on the other hand, we know that the theory of classical hydrodynamics is *actually valid for those which do.* Fluids which are "like" water and air are now called Newtonian fluids. They are differentiated from plastics, pseudoplastics, and other kinds of fluids. This differentiation is based upon the empirically different behavior exhibited by those fluids within an arrangement called "Couette-stream," or "shearing stream." The fluid under consideration is enclosed between two planes which can be moved in relation to each other. The fluid is affected by the motion of the planes, and in time there will be a certain distribution of velocity. As such, there is an empirically determinable relation between the shearing stress and the velocity gradient within the fluid. Newtonian fluids are those which show a direct proportionality between the shearing stress and the velocity gradient.

Now in our case, the point worthy of notice is that this linear relation between the shearing stress and the velocity gradient in the Couette-stream already plays an important role in the formulation of the theory of classical hydrodynamics. The principles of continuity and energy do not contain anything germane to fluids or even to a certain kind of fluid. But the Navier-Stokes equations can only be obtained by presupposing the proportionality between shearing stress and velocity gradient characteristic of Newtonian fluids.

Presupposing that one deals with a mechanical system, one sets up equations for the three components of the impulse (of an infinitesimal fluid volume) accounting for external forces (X), for differences in pressure ($\delta p/\delta x$), and for the internal friction caused by the viscosity of the fluid. The latter implies the problem itself because it is set up as a three dimensional tensor:

Can Science Reach Truth?

$$Y = \begin{bmatrix} \sigma_{xx} & \sigma_{xy} & \sigma_{xz} \\ \sigma_{yx} & \sigma_{yy} & \sigma_{yz} \\ \sigma_{zx} & \sigma_{zy} & \sigma_{zz} \end{bmatrix}$$

whose nine coefficients upon being partially differentiated yield the contributions corresponding to the changes in impulse:

$$\rho \frac{du}{dt} + \rho \cdot X - \frac{\delta p}{\delta x} + \frac{\delta \sigma_{xx}}{\delta x} + \frac{\delta \sigma_{yx}}{\delta y} + \frac{\delta \sigma_{zx}}{\delta z}.$$

$$\rho \frac{dv}{dt} + \rho \cdot Y - \frac{\delta p}{\delta y} + \frac{\delta \sigma_{xy}}{\delta x} + \frac{\delta \sigma_{yy}}{\delta y} + \frac{\delta \sigma_{zy}}{\delta z}.$$

$$\rho \frac{dw}{dt} + \rho \cdot Z - \frac{\delta p}{\delta z} + \frac{\delta \sigma_{xz}}{\delta x} + \frac{\delta \sigma_{yz}}{\delta y} + \frac{\delta \sigma_{zz}}{\delta z}.$$

This is a very general setup and conceivably has little empirical content, but it can be reduced to form the Navier-Stokes equations precisely by presupposing a proportionality between shearing stress and velocity gradient. In that case, the contribution of internal friction can be expressed by the second derivatives of the velocity components:[10]

$$\rho \frac{du}{dt} = \rho \cdot X - \frac{\delta p}{\delta x} + Y \left[\frac{\delta^2 u}{\delta x^2} + \frac{\delta^2 u}{\delta y^2} + \frac{\delta^2 u}{\delta z^2} \right].$$

$$\rho \frac{dv}{dt} = \rho \cdot Y - \frac{\delta p}{\delta y} + Y \left[\frac{\delta^2 v}{\delta x^2} + \frac{\delta^2 v}{\delta y^2} + \frac{\delta^2 v}{\delta z^2} \right].$$

$$\rho \frac{dw}{dt} = \rho \cdot Z - \frac{\delta p}{\delta z} + Y \left[\frac{\delta^2 w}{\delta x^2} + \frac{\delta^2 w}{\delta y^2} + \frac{\delta^2 w}{\delta z^2} \right].$$

(these are the special Navier-Stokes equations, which in addition presuppose that the viscosity does not depend on temperature).

The details of this example might interest only the mathematical physicist. As regards the general question of truth in scientific theories, this example should teach us that an empirical characteristic of a certain physical object might make its way into the structure of the related theory. Consequently, the theory is necessarily valid for that class of objects. Thus we have exactly the kind of argument required for the truth of scientific theories. The validity of the theory of classical hydrodynamics for Newtonian fluids is shown by the fact that it structurally implies a relation defining this type of fluid.

This result might be somewhat surprising, necessitating some further comments. The theory of classical hydrodynamics seems to have become trivial. I would like to emphasize that it is not trivial, because beforehand one knows only that the theory covers the possible behavior of Newtonian fluids. But many descriptions or prescriptions (for the building of apparatus) implied by the theory will be known only if one is able to integrate the questions of the theory. As such, I would reject the reproach of triviality. On the other hand, there is a sense in which I agree with the charge, but I deem it to be a virtue of the theory: True theories must be trivial—namely, in the sense that there is no risk of falsification in the application of the theory. Under Popper's influence it has become a commonplace that to be risky is a virtue of a theory. But the boldness associated with a theory simply belongs to the intention of applying it to objects for which one is not sure whether the theory is valid or not. A "bold theory" is a conjecture; a true theory is not.

With my last remark we have already embarked upon a discussion of our achievements thus far. I would like to elaborate further so as to ensure an adequate understanding of this matter.

My first point is a restrictive one, for our results should not be overestimated. I have given one example in which the necessary validity of a theory for a certain class of objects is understandable. Another could be given by reconstructing Kant's argument for the validity of Newton's theory. Although both these examples exhibit some necessary relations, they also reveal important differences in the respective structures of the arguments. Kant's argument is transcendental; i.e., it proceeds from the conditions of experience. The other argument is based on "species of objects" in nature. There might be other kinds of arguments for the validity of theories in natural science. The field is large, and the subject is complex. Much more research must be carried out in order to obtain a more complete insight into what scientific truth may be. What I propose is but an idea that may encourage such research. The example shows that it is possible to have more than merely a good conjecture about nature, and it exemplifies what the validity of a theory might mean. Thus it considers science, not as a means for effective betting, but as a means for construction, and it takes the practical and technological side of science seriously.

The central point of the example from classical hydrodynamics is that it is possible to talk about species not only in the realm of the life sciences. Kant's theory has no room for that. It was therefore wise to provide an additional example. The "existence" of species of objects is neither new nor surprising. Science has always dealt with electrons and iron, as it did with horses. As such, the scientific world is not only what is the case (*"Die Welt ist alles, was der Fall ist,"* Wittgenstein), but it also contains species of objects. Without doubt this is the point where far-reaching questions must be raised by a philosophy of nature. From our example we can learn the following: The specificity of natural objects is not simply revealed by nature itself. The specification of liquids could be performed in a different way, e.g., in accordance with colors. That liquids are specified on the basis of their behavior within the setup of a shearing stream belongs to fluid mechanics. Therefore, if we must say that an insight into what is essential for a certain type of object is required to reach truth in science, this requirement can be reduced to what is essential for the practical purposes of a particular scientific discipline. Moreover, one may add that the technical equipment of a discipline is used to make sure that nature reveals itself "typically."

My last remark concerns the value of scientific truth itself. I consider it to be a good thing that on the basis of our considerations it again makes sense to speak of truth in science. This does not mean, however, that science is thus able to fulfill all the expectations that are usually connected with an emphatic idea of truth. Husserl's arguments in *Die Krisis der europäischen Wissenschaft*[11] are still valid. If we can reach truth in science, that is not to be taken as *the truth*. One should rather speak of a part of truth. Albeit true knowledge, scientific knowledge continues to be piecemeal knowledge. This also means that scientific truths have their history and their preconditions. It could easily have been objected that the preceding argument presupposed the specification of fluids on the empirical side and the conceptualization of mechanical objects in general on the theoretical side. That is true. The argument for truth merely proves that both these representations of objects fit within classical hydrodynamics. Thus we must keep in mind that scientific truth is always contextual truth. That is the reason for its historicity; the contexts of science develop historically and they may change. The truth, however, does not change; whenever nature is thematized within the same context, one arrives at the same results.

Finally, one should keep in mind what can be said about kinds of knowledge, types of science, and their dialectics.[12] Our science is a specific one, and it has a specific relation to objects. As such, its truth, when isolated, may also turn into untruth (in the sense of the Greek word "pseudos"). Just as science is a particular way of disclosing nature, it is at the same time a way of concealing it.

Notes

This chapter was presented in German at the University of Göttingen on May 25th, 1984. It has been discussed with Prof. H. Törnebohm (Göteborg) and the physicist W. Kutschmann (Darmstadt), to whom I am indebted for some fruitful comments.

This chapter first appeared in the *Graduate Faculty Philosophy Journal* 12 (1987): 25–34. © 1987. Reprinted by permission.

[1.] L. Laudan, *Progress and Its Problems: Towards a Theory of Scientific Growth* (London: Routledge and Kegan Paul, 1977), p. 224.

[2.] I deliberately make a fresh start with the concept of truth. Recent theories of truth are in most cases restricted to logical and semantic considerations. That is, they exclude epistemological and ontological questions. In this paper I am concerned with the question of truth within science, in particular within physics. This makes it necessary to take into account the particular type of knowledge that science is. A clear differentiation between the subject and the object of knowledge is one of its characteristics. For a survey of modern theories of truth, see G. Skirbekk, *Wahrheitstheorien* (Frankfurt: Suhrkamp, 1977); and L. B. Puntel, *Wahrheitstheorien in der neueren Philosophie* (Darmstadt: Wissenschaftliche Buchgesellschaft, 1978).

[3.] For the notion of alienation from nature as a precondition of science, see my paper "Kant's Epistemology as a Theory of Alienated Knowledge" (Frankfurt: Surhkamp, 1983), and H. Böhme and G. Böhme, *Das Andere der Vernunft,* in Robert E. Butts (ed.), *Kant's Philosophy of Physical Science* (Boston: Reidel, 1986), pp. 333–350.

[4.] Quite naturally the "non-statement view" of scientific theories does not allow for the question of truth.

[5.] E.g., because they cannot be integrated for those cases. I am alluding here to the problem first raised in a collective paper entitled "Finalization Revisited," in *Finalization in Science,* ed. W. Schäfer (Dordrecht/Boston: Reidel, 1983), addressing the question of what it means to say that a theory is valid for an object but cannot be applied to it.

[6.] J. Habermas, "Wahrheitstheorien," in *Wirklichkeit und Reflexion, W. Schultz zum 60. Geburtstag* (Pfullingen: Neske, 1973), pp. 211–265.

[7.] The motto of the Royal Society, "nullius in verba," characterizes the common understanding of science. But at very important points of the history of science, scientists had nothing but verbal arguments to offer, e.g., Galileo for his law of free fall.

[8.] G. Böhme, "Towards a Reconstruction of Kant's Epistemology and Theory of Science," op. cit. This article also provides more details on the interpretation of Kant that undergirds the few remarks made here.

[9.] A more extended analysis of this case is given in my paper "Autonomization and Finalization," in *Finalization in Science,* op. cit., and in G. Böhme, "On the Possibility of 'Closed Theories,'" *Studies in the History and Philosophy of Science,* vol. 11, 1980, pp. 163–172.

[10.] For the mathematical details, see K. Oswatitsch, "Physikalische Grundlagen der Strömungslehre," in *Handbuch der Physik,* ed. Flügge, vol. VIII, 1, pp. 1–124 (Berlin/Göttingen/Heidelberg: Springer Verlag, 1959).

[11.] E. Husserl, *Die Krisis der europäischen Wissenschaft und die Idee der transzendentalen Phänomenologie* (The Hague: Martinus Nijhoff, 1962).

[12.] G. Böhme, *Alternativen der Wissenschaft* (Frankfurt: Suhrkamp, 1980).

5
Science and Other Types of Knowledge

"As a scientist I cannot but be a democrat, for the realization of the demands resulting from the laws of Nature, from the nature of man, is only possible within a democratic state." This is a quotation from Rudolf Virchow, writing to his father shortly after the 1848 revolution. (In German: "Als Naturforscher kann ich nur Republikaner sein, denn die Verwirklichung der Forderungen, welche die Naturgesetze bedingen, welche aus der Natur des Menschen hervorgehen, ist nur in der republikanischen Staatsform wirklich ausführbar.")[1] This close, even essential, connection between science and democracy sounds strange to our ears; we feel a lack of evidence. More than that: Political values such as "democracy," "freedom," and "autonomy" are called up as a background for science criticism; and science on the contrary, is blamed for being "imperialistic," "hierarchical," "oppressive," and "disabling."[2] This type of science criticism is related to the so-called crisis of acceptance, which is mainly an outcome of the destructive consequences of the scientific-technological progress. But it is clearly different as well. One can emphasize this difference by saying that this type of criticism would hold even if science and technology did not have destructive consequences. Then people would say, "We do not want this kind of goods, or we don't want them this way." For the "scientification" of the lifeworld is connected with the loss of people's abilty to help themselves; they become dependent on experts. "Verwissenschaftlichung" is connected with bureaucratic centralization, with increased governmental power, with a loss of regional differences and personal knowledge.

It is within this context of criticism that we realize that the "nonscientific traditions in higher education" constitute a problem. We feel that the institutional incorporation of nonscientific knowledge traditions into the university system may result in the loss of their peculiarity and a further reinforcement of the imperialism of science. But if we want to enter a plea for nonscientific knowledge traditions, if we want to preserve and improve their particular character, we must first perform a very hard task; that is, we must demonstrate that they are something peculiar at all. For the general argument of the "scientistic promoter" is that science is just *better* than any other type of knowledge. This "being better" has a double meaning. On the one hand it means that other types of knowledge are, in principle, not different from science, that science is just more differentiated, more systematic, more elaborated. On the other hand, it means that science is good for all, that it has no specific function; or, the other way round, that types of knowledge do not differ by their functionality. To get clear about these points will be our main task, for this is the presupposition of all pragmatic considerations, including knowledge policy.

But we must turn back to the relationship between science and democracy. We must recall the primordial interrelation of science and republican ideas and the gradual divorce of the scientific movement from the democratic. That will lead us to one of the main motives for scientification—namely, the development of professions, or the tendency toward the professionalization of occupations. We shall then be in a position to say, in a sociological context, what the scientification of nonscientific types of knowledge means. And, finally, we shall return to the task of determining the difference between science and other types of knowledge.

Science and Democracy

It is well known that our type of science—in contrast to medieval scholastic science or Greek science—is a product of the emerging bourgeois class. Thus it shares some common ideals with the bourgeois class, which, in the beginning, was anti-authoritarian, dedicated to social and political progress, bound to produce *public* knowledge; it was egalitarian and so would produce knowledge that was universal, i.e., valid for everybody. These ideals of the new learning of the seventeenth century were directed against the hermetic, authoritarian, elitist knowledge of the schools. Science was revolutionary in its early days. It produced knowledge of use and accessible to everybody. Today, by contrast, it is sufficient just to enumerate these ideals to recall how far the actual practice of science stands off from them. Nevertheless, even in our century, sociologists of science such as Robert Merton (1968) have characterized the normative structure of scientific practice as being the prototype of liberal democracy. Thus, for example, Merton's norm of universalism implied that the value of scientific results should be independent of the race, sex, and social background of their producer; thus, too, the career and reputation of a scientist had to be independent of these personal and social backgrounds. According to the norm of communism or communality, scientific results had to be open to the public, so publication was an essential part of the scientific production itself. The norm of organized skepticism meant some sort of continuous revolution—that is, the steady openness of established ideas for revision. And, to mention the last of Merton's norms in his set of scientific ethics, the norm of disinterestedness postulated the independence of scientific products from the private interest (i.e., objectivity) of its producers.

Considering these ideals, one might say that what we have is just the usual difference between ideal norms and actual practice. But to say this is not adequate to explain what is problematic with the democratic sense of science. In fact, the norms mentioned have been fulfilled, more or less—but only *within* the scientific community, not at the borderlines[3] or among the general public. What we have to face is the fact that science soon after its beginnings has divorced itself from the general democratic and emanicipative movements. This process has been analyzed in detail by Webster (1976), Mendelsohn (1977), and my colleague v.d. Daele (1977). I refer here to the process of the institutionalization of the new type of science within the Academies toward the end of the seventeenth century. One fact may highlight the paradoxical character of this process: Whereas freedom of speech and publication is one of the general bourgeois demands, the scientific societies received the *privilege* of censureless publication from the restorative government of Charles II. The point is that modern science was institutionalized as a new kind of corporation with particular privileges, with certain thresholds of access, and with relations to the outside that differed widely from the internal relations among members. So, science as a social structure put to work the ideals of liberal democracy largely within its own house—but its relations with the outside grew more and more hierarchical and mystical; they became relations of domination. In a bad humor one might compare the scientific community to the *libertins:* democratic and fair to members, but without scruple toward the outside.

Let us consider some of the main steps of this process. In its early times, modern science was performed by amateurs. It had no definite relation to the universities, and no threshold of access. Scientific journals were addressed to the learned, to the curious.[4] During the latter part of the eighteenth century, specialized journals were founded, and the new science was introduced into the universities. This was an important event, inasmuch as the universities, since their beginnings in the twelfth and thirteenth centuries, gave access to higher occupations such as physicians, lawyers, and divines. (These were the original occupations or professions.) The introduction of modern science into the university resulted in the professionalization of science and the scientification of university learning. Science itself became a profession—that is, a job that cannot be performed without highly sophisticated training, to which one gains access only through university exams and certificates. At the same time, science became the standard type of knowledge within the university setting, such that even the old departments of medicine, law, and theology had to accommodate to it and prove that their procedures were "scientific." Beginning with the eighteenth century, but characteristic of the nineteenth century as well was a process that can be called a "run to the universities"—that is, a continuous striving of other occupations to become professions. Well known is the professionalization of such medical occupations as surgeon, barber, and apothecary, which became academic occupations.[5] But the same is true of engineering[6] and almost all teaching jobs. We turn now to a consideration of the reasons for this striving toward professionalization and the meaning of the related scientification of the traditions concerned.

Professionalization and Scientification

Regarding the topic of nonscientific traditions in higher education, it might be true, from an administrative point of view, that the incorporation of other traditions such as nursing, teaching, and the fine arts into the university system is a question of quantities and the organization of social goods and demands. But we should not forget that the practitioners of these traditions have some interest in this incorporation—in short, an interest in professionalization. So we don't have a process that is forced upon a certain group of people whom we now have to defend against administrative suppression; the problem, sad to say, is much more complicated than that. Indeed, professionalization may be against the "better understood" interests of the traditions concerned.

What are the reasons for the general striving for professionalization?[7] Recall the 100-year struggle of engineers to get the right of doctoration and to transform the institutes of technology into universities. This was a struggle for social recognition, independence, and material rewards. As for recognition, with the disintegration of the feudal hierarchies, academic learning became the basis of a new nobility within the bourgeois class.[8] In bourgeois society, learning was equated with the principle of achieving social status, which in turn paralleled another principle—namely, property. Within this competition—learning versus property—learning had the advantage of relation to state occupations. Public appointments were originally accessible only to noblemen; the exceptions were, as ever, the traditional professions! The more the public influence of the nobility was relegated to the background, the more the access to public appointments and thus to influence in social and political affairs was through academic learning. This principle—namely, distribution of social opportunities according to level of certified education—was increasingly extended. The reason for this was the extension of the administrative sector itself, the bureaucratization of the big companies, and the relative decrease of the productive sector relative to the increase of the distribution and services sector. As a consequence, according to R. Bahro (1977) and Konrad and Szélenyi (1978), we can speak about domination by the intellectuals in our society. In this connection, the American sociologist A. Gouldner (1980) has suggested that knowledge is a new sort of capital and that the intellectuals are going to be a new dominating class. We need not examine this thesis in detail, but we do have to take into account what it adds to our discussion thus far: that, in addition to the distribution of social opportunities according to the degree of education, there is a fundamental antagonism between the "educated" men and women in our societies and the noneducated—that is, between the participants of higher education and the rest.

We have thus seen a very short sketch of the social history of academic learning. It began with privileges for a very small corporation of amateurs, and it ended with the recent tendency of the intellectuals to form a dominant class in our society. Today, almost all leading positions in economics, production, administration, and bureaucracy are inaccessible without an academic degree. Hence political and social power falls to the academics. The consequences concerning the social behavior toward higher education are as follows: The participants of higher education try to reproduce themselves on the basis of higher education and, at the same time, try to keep the threshold of access to higher education as high as possible. They try to establish a hierarchical relationship to other types of knowledge and their bearers. And these bearers of other knowledge try to participate in higher education in order to transform their own traditions into higher education.

In the next section we shall consider what this transformation of knowledge traditions means: namely, scientification. For the moment, however, we have to discuss scientification merely on a sociological level: The professionalization of certain occupations and social practices means the scientification of the respective type of knowledge. We can now state that the scientification of certain social contexts is an interest of a particular group within society—specifically, the members of the intellectual class and the bearers of higher education. This process extends the range of their responsibility and, hence, their power. If a problem within society is defined on a scientific level and in scientific terms, then it is within the responsibility of the scientifically trained staff to solve it. In this way, problems of education and social welfare become problems for professional social workers and psychologists. Thus scientification of social practice means the extension of social influence for the bearers of higher education. It also means autonomy for the average man and woman in society. Whether this shift in responsibility is an advantage for the clients thus produced, given the excellence of scientific knowledge, is a question to which an answer will not be possible until we have an idea of the difference between scientific and nonscientific knowledge.

Differences Between Types of Knowledge

Let us now approach the question as to how science and nonscientific types of knowledge differ. Of course, this is a nearly unsolvable problem. In order to determine the peculiarities of the different types of knowledge, we have to study each type in particular; we can't do it on a general level. Thus I have to transfer the problem to each of you inasmuch as you are bearer and practitioner of a certain type of nonscientific knowledge. I have researched such differences—in the context of knowledge of midwifery[9] and everyday-life bodily competences,[10] so I have some experience in such questions. What I can do on a general level is to indicate within which dimensions and in which respects the differences between science and nonscientific knowledge are to be sought. The problem here is that in doing so in general I am obliged to delineate the dimensions on the basis of a characterization of *science,* thus reproducing the domination I want to resist on a theoretical level. But let me first summarize what it is, according to the foregoing analysis, that we want to know.

Science is the dominant type of knowledge. This fact is legitimated by the claim that science is better than any other type of knowledge. So we want to know whether the relation of science to other types of knowledge is a difference to be determined on some scale of eminence or function. Science has turned out to be the property of a certain group of persons—namely, the scientific community and intellectuals—who exercise domination over the nonscientists, their clients. What, then, is the relationship between types of knowledge and the people concerned? In consideration of these questions we cannot hope to find an answer if we are looking for the difference between types of knowledge on a merely cognitive level. So, what is knowledge?

It is quite common to think that knowledge is what is written in a book and can be learned. But this position implies a reifying concept of knowledge. What is written in books, what is stored in retrieval systems, even what is contained in the heads of people is the content of knowledge. Knowledge itself has to be differentiated from "knowing"—that is, from the way in which people take part in the things to be known.

Things to be known, taken as a totality, are the intellectual or cultural wealth of a society. So knowing is to be defined as the way of participation in the cultural wealth of a society. Seen from this point of view, social opportunities are distributed not only according to property, the material wealth of the society, but also according to knowledge. On the basis of this definition, knowledge as participation in the cultural wealth of society, we can say that the differences between types of knowledge have to be determined mainly with respect to the type of participation involved. The differences of content will largely turn out to be dependent on the way of participation. This, in philosophical terms, is the thesis of constitution theory—that is, the claim that, for instance, what can be known about nature is dependent on the way of approaching nature.

The possible differences in participation can be broken down along six dimensions, following Parson's pattern variables:[11] Knowledge can be (1) personal or impersonal, (2) conservative or progressive, (3) diffuse or specific, (4) implicit or explicit, (5) particular or universal, and (6) empathetic or dominant.

Personal Versus Impersonal

This variable is one of the most important to a characterization of modern science, in contrast to the traditional science it was to overcome and in contrast to nonscientific types of knowledge today. First, consider the way one gets experience. Given the success of modern science, the experience you rely on need not be your own. As the norm of reproducibility and the organization of data production according to rules warrant the validity of experience for everybody, you can build your own research on the basis of others' experience. In other words, you need not be "experienced" yourself. The situation is very different, however, with respect to the knowledge required for personal care or handicraft. A good example is Chinese medicine, in contrast to European medicine: In the case of acupuncture, you need to have tried each practice on your own body to be competent. As a second case in point, let us consider the type of knowledge tradition involved. Here, in principle, scientific learning is independent of the personal relationship between teacher and pupil. It is true that even natural science requires schooling, and the institution in which one has learned science may make a difference. But even science cannot do without some of the more traditional ways of practice:[12] The ideal is the other way around. This is clearly shown by the fact that scientific knowledge is published and thus can be arrived at through books; hence scientific experience should be achieved according to explicit rules. By contrast, learning in other traditions (e.g., in the arts) is supposed to be based on a personal relationship between teacher and pupil.

As a third point within this dimension I would like to mention the relationship between knowing and personal maturity. It was one of the main steps in the development of modern science to deviate from the old philosophical idea that one can become wise only through a process of becoming mature as a man or woman. Scientific competence is independent from moral status; and, conversely, being a good scientist does not contribute to the quality of one's personality. Other types of knowledge are different in this respect. Consider this example, which I have studied in detail: Traditional midwifery presupposed that the woman involved was a 'mature' woman; but modern scientific midwifery can be performed by any well-trained 18-year-old woman.

Conservative Versus Progressive

I confess that I hesitated in my use of these terms, because in trying to defend nonscientific types of knowledge it is difficult to say that they are nonprogressive. So let us take these terms in a descriptive, nonpolitical sense: Science is progressive inasmuch as being a scientist means being a researcher. The development of modern science was the overcoming of the old ideal of the learned by the new ideal of the researcher.[13] To participate in scientific knowledge means to look for something new. So scientific knowledge is undergoing continuous change; innovation is more than reproduction. As a consequence, scientific knowledge is essentially without history; that is, history does not belong to the corpus of scientific discipline. This premium on innovation was one of the main reasons for which nonscientific types of knowledge were defeated by science. The fact that science was always able to present its own contributions as the very latest ones seemed to recommend science as the best type of knowledge. That this is a very dubious conclusion may be supposed, given that the field in question has a certain historical invariance. This is the case with personal care in childbirth, illness, and death. But in fact these areas, too, are undergoing continuous change—because the scientific or professional way of knowing about them is.

Diffuseness Versus Specificity

This is the dimension commonly referred to when one is about to express some uneasiness about scientific knowledge. One says that other types of knowledge are more "intuitive," and that science gets away from the "whole." In fact, the effectiveness of scientific exactness and distinctness is based upon a differentiation of variables in one field and by the disciplinary partialization among fields. Thus, with the scientification of knowledge, problems often fall between the disciplines and the integrative competences of men and women are not exercised. Consider scientific medicine, for example. Here the rule is: The progress of scientification is connected with specialization. And this is not always beneficial to the patient. So one characteristic of nonscientific types of knowledge is diffuseness. This diffuseness is a handicap with respect to the competition with science, but it is often an advantage for the people concerned.

Implicit Versus Explicit

This variable could also have been called "practical versus theoretical." The point is that knowledge as participation in the cultural wealth of a society includes types of knowledge that entail practical competences and skills. The question is whether knowledge is made explicit (e.g., by giving rules or writing something down) or is implied in one's behavior. The process by which knowledge becomes more theory oriented seems to be a consequence of its explicitness. In this way knowledge can become a "Sinnprovinz" of its own; that is, it has a field of meanings and problems above concrete social action. Scientification is generally connected with explication and theorizing. Thus, for example, artisanship and technology consist of practical skills; their products are things, not papers. But at the moment when technology started to become scientific, it was theorized and papers came to be its products. This variable—implicitness versus explicitness—must be scrutinized thoroughly, because the common supposition is that explication means improvement of knowledge. Actually, a dichotomy into theory and practice may result, thus forming a relationship of hierarchy as in the classic case between the architect and the bricklayer.

Particularity Versus Universality

This, too, is a category in which nonscientific types of knowledge seem to be defeated from the first moment on. Science produces knowledge, which is valid all over the world, for each person, in each place. So the old prescription of artisanship to travel seems to be an expression of the regionality of its knowledge. But recently we have learned to appreciate regionality. On the one hand, we have noticed what is implied when a certain knowledge is adapted to the regional conditions (as in the case of adapted technology). On the other hand, we have learned to understand what is meant by the universality of science: not that science is valid under *all* conditions, but that it is valid under definite conditions. Thus we must establish adequate conditions before introducing our technology into Third World countries. So a type of knowledge that is adapted to particular conditions and hence has only a limited validity may be superior to universal knowledge.

Empathy Versus Domination

This last dimension serves to define the difference between science and nonscientific types of knowledge. Remember that knowledge is the relationship of participation to what is known. At first glance this seems to be of interest only to the Platonist, if one is considering the claim of Thales or other mathematical examples. But if the object of knowledge is nature or the other man or woman, it makes a great difference whether participation is empathetic or a relation of dominance. Science is a type of knowledge that is based on a fundamental distance between the subject and the object known. This distance is generally formulated as the postulate of objectivity: The object of knowledge should not be influenced by the subject of knowledge. But more important, the subject of knowledge should not be influenced by the object of knowledge. This conclusion could not have been drawn before the social sciences emerged, because the empathy with nature was driven out of science in the early seventeenth century. So coldness and mere instrumental behavior toward nature seem now to be natural. This stance may be criticized today as far as the natural sciences are concerned, but we have to realize that such nonscientific types of knowledge as those involved in teaching, nursing, gardening, and farming presuppose a good amount of empathy with nature. Yet empathy is to be differentiated from domination. Knowledge through empathy means that the subject of knowledge is affected by, or suffers because of, its object. This is exactly the contrary to domination. Domination means that the object may be determined by the subject of knowledge, and not vice versa.

In discussing these pattern-variables, I have mentioned several examples of nonscientific knowledge—namely, nursing and personal care, other aspects of medicine, artisanship and the arts, technology and teaching, farming and gardening. Let me now sum up the characteristics in one example that I have studied more closely: midwifery, as distinct from scientific obstetrics. Traditional midwifery was based on knowledge that had been transferred in personal apprenticeship relations. It presupposed personal maturity and experience; that is, the midwife was supposed to have had children of her own. And it encompassed an invariant, only slowly changing corpus of knowledge, which was the basis of a rather diffuse, holistic competence of personal care. Midwifery was a competence integrated into the lifeworld and, hence, was affected by local customs and adapted to regional social conditions. It was an empathetic competence and thus could be performed only by women; yet it was not a competence taken up with guiding and controlling the process of childbirth. By contrast, scientific obstetrics is a rather impersonal job, based on technical competence. It can be learned in schools and through books. It presupposes neither a certain level of personal maturity nor biographic experience. A highly specialized competence, it represents universal knowledge in the sense that it is valid all over the world (at least where the appropriate clinical equipment is supplied). And it is not affected by empathy; indeed, empathy would make impossible the optimal technical guidance of the process of birth. I leave it to you to suggest other examples. What should have become clear here is that nonscientific types of knowledge differ from science in degree but also that there are structural differences. On the basis of these nonscientific types of knowledge the domination of science over other types of knowledge cannot be legitimized. Conversely, we see that the scientification of social practices and other types of knowledge will have very far-reaching consequences. The obsolescence of these other types of knowledge will not be balanced by the advantage offered by science. Needs will go unmet, problems unsolved. Nonscientific types of knowledge are competences that have social functions different from those of science.

Summary

Let me summarize this chapter with three theses:

1. Science, by its very nature, implies some general democratic values. But these are applied only within the internal communications of the scientific community. Science as a social factor developed from a small privileged corporation into a sort of ruling class of experts.
2. The striving for professionalization of nonscientific traditions has resulted in their transformation into scientific knowledge. The reason for this is not the superiority of science over other types of knowledge but the fact that in our society social opportunities are largely distributed according to certificates of knowledge.
3. Science and nonscientific types of knowledge differ structurally. These differences form the basis of an alternate functionality. Whether science or nonscientific knowledge is required depends on whether a personal, empathetic relationship to the object concerned is actually necessary, whether one actually needs a rather particular and regionally adapted competence, and whether a more holistic or more specified approach is adequate.

In conclusion, then, I suggest that we can become more impartial toward nonscientific traditions if and only if we can cope with the social motives underlying professionalization.

Notes

1. Letter from May 1, 1848 (Virchow 1907).
2. Böhme and v. Engelhardt 1979; Böhme 1979; Illich 1978; Bergendal 1981.
3. In my view, Mulkay's (1969) criticism that norms of science do not "exist" is in fact concerned with a borderline affair (the Velicovscy case).
4. Kronick 1962.
5. Ackerknecht and Fischer-Homberger 1977.
6. Manegold 1970.
7. For elaboration of this concept, see Hesse 1968; Parsons 1968.
8. Paulsen 1966; Schnabel 1948–1951, Vol. 3.
9. "Wissenschaftliches und lebensweltliches Wissen am Beispiel der Verwissenschaftlichung der Geburtshilfe," in Böhme 1980.
10. "Die Verwissenschaftlichung der Erfahrung," in Böhme and v. Engelhardt 1979.
11. I think it is legitimate to proceed on the basis of Parsons' variables; indeed, they were developed in order to characterize modern versus traditional social structures. But we run the risk of inheriting through these variables Parson's position that "modern" is better in any case.
12. This point has been emphasized by the conservative sociologist of science, M. Polanyi (1962). But clearly the traditional pattern need not be essential to this point. Accordingly, Collins (1974), who has done some empirical research on such questions, demonstrates that the more handicraft-like traits within scientific practice are prevalent during the *initial* phase of development.
13. Znaniecki 1968.

Bibliography

Ackerknecht, E. H., and Fischer-Homberger, E. "Five made it, One not—The Rise of Medical Craftsmen to Academic Status during the 19th Century." *Clio Medica,* Vol. 12 (1977): 255–267.

Bahro, R. *Die Alternative.* Frankfurt: EVA 1977.

Bergendal, G. "Higher Education and Knowledge Policy," unpublished paper (Malmö, September 1981).

Böhme, G. *Alternativen der Wissenschaft.* Frankfurt: Suhrkamp, 1980.

———. "Die Entfremdung der Wissenschaft und ihre gesellschaftliche Aneignung." *Wechselwirkung,* No. 3 (1979).

Böhme, G., and v. Engelhardt, M. *Entfremdete Wissenschaft.* Frankfurt: Suhrkamp, 1979.

Collins, H. M. "The TEA set: Tacit knowledge and scientific networks." *Science Studies,* vol. 4 (1974): 165–186.

v.d. Daele, W. "Die soziale Konstruktion der Wissenschaft." In G. Böhme, W.v.d. Daele, and W. Krohn, *Experimentelle Philosophie. Ursprünge autonomer Wissenschaftsentwicklung.* Frankfurt: Suhrkamp, 1977.

Feyerabend, P. *Erkenntnis für freie Menschen.* Frankfurt: Suhrkamp, 1979.

Gouldner, A. W. *Die Intelligenz als neue Klasse.* Frankfurt: Campus, 1980.

Gurvitch, G. *The Social Frameworks of Knowledge*. London: Harper Torchbooks, 1971.
Hesse, H. A. *Berufe im Wandel. Ein Beitrag zum Problem der Professionalisierung*. Stuttgart: Enke, 1968.
Illich, I. u.a. *Entmündigung durch Experten*. Reinbeck: Rowohlt, 1978.
Konrád, G, and Szélenyi, I. *Die Intelligenz auf dem Weg zur Klassenmacht*. Frankfurt: Suhrkamp, 1978.
Kronick, B. A. *A History of Scientific and Technical Periodicals: The Origins and Development of the Scientific and Technological Press 1665–1790*. New York: Scarecrow Press, 1962.
Manegold, K. H. *Universität, Technische Hochschule und Industrie*. Berlin: Duncker u. Humblot, 1970.
―――. "Die Entwicklung der technischen Hochschule Hannover zur wissenschaftlichen Hochschule. Ein Beitrag zum Thema Verwissenschaftlichung der Technik im 19. Jahrh." *Verein dt. Ingenieure (Hrsg.) Technik in Einzeldarstellungen,* Vol. 16 (1970): 13–46.
Mendelsohn, E. "The Social Construction of Scientific Knowledge." In E. Mendelsohn, P. Weingart, and R. Whitley (eds.), *The Social Production of Scientific Knowledge,* pp. 3–26. Dordrecht and Boston: Reidel, 1977.
Merton, R. K. "Science and the Social Order." In R. K. Merton (ed.), *Social Theory and Social Structure,* pp. 591–603. London: Macmillan, 1968.
Mulkay, M. "Some Aspects of Cultural Growth in the Natural Sciences." *Social Research,* Vol. 36, No. 1 (1969): 22–52.
Parsons, T. "Die akademischen Berufe und die Sozialstruktur." In T. Parson, *Beiträge zur soziologischen Theorie* (edited by D. Rüschemeyer), pp. 160–179. Neuwied a. Ph.: Luchterhand, 2. Aufl. 1968.
Paulsen, F. *Die deutschen Universitäten und das Universitätsstudium*. Hildesheim: Olms, 1966 (1902).
Polanyi, M. *Personal Knowledge*. New York: Harper Torchbooks, 1962.
Schnabel, F. *Deutsche Geschichte im 19. Jahrhundert*. 4 vols. Freiburg: Herder, 1948–1951.
Virchow, R. *Briefe an seine Eltern 1839–1864* (edited by von M. Rabel geb. Virchow). Leipzig: Engelmann, 1907.
Webster, C. *The Great Instauration: Science, Medicine and Reform 1626–1660*. New York: Holmes and Meier, 1976.
Znaniecki, F. *The Social Role of the Man of Knowledge*. New York: Harper Torchbooks, 1968.

6

Science for War and Peace

Useful Science

I proceed on the assumption that the existence of the atomic bomb represents a major offense to science, a painful wound that afflicts every individual scientist more or less consciously. As a scientist one believes that one's own activity of pursuing science is something good, useful, and reasonable. One has to believe this because the extraordinary amount of self-denial and self-discipline required by science could not otherwise be sustained. The atomic bomb and everything it symbolically encompasses—from chemical weapons to the neutron bomb—has fundamentally shaken this belief. And the individual cannot bear this disappointment. The order of the day is suppression of and flight from this unpleasant fact. The most common procedure is one of separation: One separates the questionable part of scientific activity from one's own. One distinguishes between basic and applied science, between science and technology, between the natural and the social sciences. Finally, one distinguishes between science and politics. It is always the others who bear the responsibility so that the individual himself can in any case continue to work undisturbed. In contrast to that, my first maxim would be to view the problem as a collective one, as a problem that commonly concerns the bearers of scientific-technological progress.

Some historical illustrations follow. What we understand by science is a product of the Renaissance. Viewed socio-historically, it emerged as a coming together of the traditions of scholars on the one hand and craftsmen on the other. This is the source from which the fundamental methodological unity of science and technology originated. In addition, science is a product of the emerging bourgeoisie. Its innate public character and its claim of universal validity sprung from there. Science is a collective undertaking; i.e., the individual as scientist is not a scholar but someone who makes a contribution to the process of science. Science could also have been understood in a different manner in the past: as inspirational knowledge for the individual, as orientational knowledge in the cosmic order. Different normative orientations apply for modern science. They were formulated by Francis Bacon, the great ideologist of the new science. Scientists should use the talent of their reason "for the benefit and prosperity of humanity." Social progress, improvement of living conditions, fighting disease are its tasks. Science should be useful. However, for Bacon, a science that permitted the development of gunpowder was as useful as one that made possible the conservation of food. Following this understanding of social usefulness, Galileo commended the telescope as an instrument of war, Haber developed ammonia synthesis and poison gas, and Werner von Braun created the V2. Thus, social progress has also always meant the improvement—the modernization—of weapons.

The development of the atomic bomb introduced a fundamental change into this comprehension of social usefulness. In the past, to be socially useful meant: useful to friends, harmful to enemies. The development of the atomic bomb was a process pursued with this understanding in mind; but it was later condemned because it rendered obsolete the differentiation that something could be harmful to the enemy but useful to the friend.

One can express it thus: The universality of science asserts itself in the scientization of war, too. The Teutons and Romans or the Christians and Turks may have confronted one another with war technologies whose differences were culturally determined; but today it is the case that every improvement in war technology will, in the short or long term, be directed against one's own position. Thus, in the wake of the atomic bomb, one must ask whether any war-related scientific activity at all can be designated as socially useful. In this sense, the bomb means the end of Bacon's program. The norms of science must be conceptualized anew.

Furthermore, the atomic bomb indicates the limits of an understanding of politics according to which (following Carl Schmitt's definition)[1] politics designates the difference between friend and enemy. In the sense of this traditional understanding of politics, war belongs in principle to the spectrum of possible political modes of action. And in this understanding of war, since the end of the last century, a change has been taking place that has found expression in the various proscriptions of war in international law. The reason for this change lies ultimately in the scientific-technological development of war. The possible destruction in war no longer stands in any relation to the possibly disputed interests between nations. The use of war as a political means is today a sign of a nation's deficient level of civilization; it is in every instance a case of barbarism. And if war as such lacks political legitimation, then scientists' war-related activity is in every instance without justification as well.

Historical Change as a Result of the Atomic Bomb?

The construction of the atomic bomb has definitively delegitimized war-related scientific activity; so one must wonder why war-related scientific work has continued. Just as one can correctly designate the atomic bomb as an end to science for war, one can also call it a beginning: War-related scientific work has been on the increase since World War II.

First, a review of the conduct of scientists vis-à-vis the atomic bomb prior to Hiroshima. Two things are remarkable: first, the speed with which even theoretical scientists and those involved in basic research recognized the war-related significance of nuclear fission; second, that it was the scientists themselves who commended the atomic bomb to governments and the military.

December 1938 saw the conclusion of Otto Hahn and Fritz Straßmann's work, which had proved the possibility of nuclear fission. The experiment did not make use of any of the large research installations as we know them in nuclear physics today and as they were already then developed by Ernest Orlando Lawrence, who constructed the cyclotron. Hahn and Straßmann had used an apparatus that could be placed on a table in a laboratory. However, by 1939, theoretical reflections on the energy released had led physicists all over the world (among them, Leo Szilard, Eugene Paul Wigner, and Edward Teller in the United States, and Carl Friedrich von Weizsäcker, Siegfried Flügge, and Werner Heisenberg in Germany) to consider the possibility of the atomic bomb. These thoughts did not remain theoretical but were immediately translated into political action. Noteworthy was Szilard's attempt to achieve a voluntary ban on publications from scientists working in this area. This suggestion, in itself revolutionary relative to customary practice in science, was intended not to prevent the development of the atomic bomb but simply to prevent its development in the hands of the Nazis. On the other side, it was Szilard in the United States who lobbied intensively in his efforts to win over the U.S. government and its military to development of the weapon. As early as March 16, 1939 (i.e., a quarter of a year after the discovery of nuclear fission), he got in touch with the U.S. Navy minister; this contact was mediated by the Nobel prize winner Enrico Fermi. When nothing much came of this, he persuaded Albert Einstein to write a letter to the president outlining the possibility of this new weapon and the danger that would arise if the Nazis were able to develop it. Roosevelt received this letter on October 11, 1939. There were similar lobbies also active in Germany in 1939. The physicists Georg Joos and Wilhelm Hanle contacted the Reich's education ministry, and the physicists Paul Harteck and Wilhelm Groth turned to the Reich's war ministry. The long-term success of these lobbies is well known. Founded in Germany was the uranium association, where the most important specialists, exempt from military service and supported by the state, conducted uranium research. This work was hampered, however, by bureaucratic disputes about competencies in the Nazi system and was only hesitantly done by those scientists who did not have the commitment to involve themselves in a "just cause." In Germany, on the one hand, there were no special investigations conducted on the development of the weapon. Just before the end of the war, the critical point (i.e., the condition in which a chain reaction maintains itself independently) was almost reached in the development of a heavy-water reactor. In the United States, on the other hand, enormous research installations were constructed for the production of enriched uranium and plutonium, and this in an economically irrational way. At the same time (i.e., even before the explosive as such could be reliably produced), a weapons laboratory was built in Los Alamos whose sole purpose was the development of the bomb. Then, in July 1945, the test bomb was successfully detonated. It cost the United States $2 billion, an amount comparable to an entire prewar budget. There is no doubt that scientists bear the responsibility for this development. The military was extremely skeptical and difficult to persuade; and the politicians understood little about the matter and even considered the physicists' ideas illusory. Here, the authority of Nobel prize winners was necessary in order to make any impression at all. For the scientists in the United States, the motivation to make great efforts in order to gain political, military, and economic support came largely from their fear of the Nazis. Evidence for this is the fact that the most important advocates of the bomb were emigrants. However, there were other motives involved as well. Consider, for example, the gigantism of Lawrence, who was primarily interested in expanding his cyclotron projects. The first attempts in Germany to acquire the funds for research were undertaken in the same manner, with the possibility that new weapons development would serve as bait.

When I say that scientists bear the responsibility for the development of the bomb because it came about on the basis of their political initiative and because their academic motivation profited from it, I am referring to scientists collectively. Scientists and technicians of every type participated in the development of the bomb: basic researchers such as Enrico Fermi, Karl Taylor Compton, Leo Szilard, and J. Robert Oppenheimer; explosives specialists such as George B. Kistiakowski; and a variety of technicians and engineers. It is indeed interesting to see how, in cases of emergency, scientific competence (even that of the esoteric theorist) could be functionalized for such an objective. In fact, the scientists and technicians did act as a collective. In all the reports coming from Los

Alamos, the extraordinary atmosphere of solidarity and the awareness of common interest among all participants can be sensed.

The development of the atomic bomb shaped the course of science in the following ways.

1. Large research installations were established. These institutions, with factory-like dimensions, were organized outside the universities and thus outside teaching. They constituted an independent factor from then on, faced with considerations of profitability and efficiency as well as with problems concerning recruitment, labor, and capacity utilization.
2. Science became a national security factor. Since World War II, the military power of nations has been a function of their level of scientific and technological development.
3. Scientists' attitude toward politics has changed fundamentally. In the course of time, as a result of their success all governments now retain scientists as advisers. Whether political power has accrued to the scientists themselves remains uncertain. But this could have been the case as early as 1939. Thus von Weizsäcker motivated his colleagues in the uranium association during the National Socialist regime by expressing the hope that they could gain influence over the system. But the politicization of scientists themselves, though it has remained incomplete, is a lasting result of this development.

When we ask about the significance of the change that the atomic bomb has brought about in the relationship between science and war, the facts reveal the idealistically derived delegitimation of military science only fragmentarily: First, the delegitimation of science for war has had no effect because military science itself has become an important social institution. Science for war no longer needs any special moral motivation. Second, the scientization of war has led to a militarization of science. And, third, the military-political contamination of science has produced a resistance movement within science, and the politicization of scientists has become an issue of conflict within science.

Science in Resistance

As long as the scientists were engaged in the construction of the atomic bomb, there were hardly any scruples about the meaning of their activity. Following the first test detonation and then the dropping of the bomb itself, however, terrified scientists were forced to reflect on their activities. The first organized resistance developed even before the bomb was dropped. It posed the question as to whether a mere demonstration of the bomb's effects could end the war against Japan so that the actual dropping of it would not be necessary at all. At an institute largely entrusted with the development of the first reactor in the United States and with the production of material for the bomb, the Metallurgical Institute in Chicago (MET), the so-called Franck Report was issued in June 1945. This report demanded that the bomb not be used militarily and that an international agreement be reached on the nonuse of all nuclear weapons. At the time, many scientists saw clearly that a nuclear arms race would ensue if the inclusion of this weapon in the military arsenal were not prohibited at the outset. Many scientists also believed that one could counter the danger by returning to the usual norm of science—namely, that of publishing everything known. Some scientists actually did publish on an individual basis, but this activity was considered a betrayal of state secrets. Others, such as Szilard and Bohr, tried to induce high-ranking politicians to establish a control authority within the framework of the newly founded United Nations, one that would supervise all work in the area of nuclear research. Then, in 1945, when a bill was introduced in the United States that on the contrary proposed classifying nuclear research as a matter of national security, the first collective threat to refuse cooperation was expressed by scientists. Sam Allison from MET stated that scientists would turn to research on the color of butterfly wings if their work were placed under military secrecy regulations. Shortly thereafter, the Chicago scientists founded the Federation of Atomic Scientists. In the succeeding years, this organization and its journal, the *Bulletin of Atomic Scientists,* have provided a forum for discussion of armaments research and the resistance against it.

In this case concrete resistance developed from what I referred to above as a contradiction: science's universalism and a national (i.e., particularistic) understanding of the use of science. However idealistic such resistance may appear, it has nevertheless continually led to political action. The public character of science and its universalism were understood by scientists to be norms that internationally bound them together through their work. It was on this basis that the so-called Pugwash Conferences were established in 1957. At these conferences, which still take place at least once a year, scientists have attempted to conciliate among politically hostile nations. And, indeed, they have been successful in their efforts to initiate various arms control negotiations and to create informal contacts among politicians.

The same year witnessed there a dramatic act of political resistance on the part of scientists in the Federal Republic of Germany (FRG). At that time the Adenauer administration had considered equipping the army with nuclear weapons. Eighteen of the most important German atomic scientists declared publicly that in their opinion a small country like the FRG could "best protect itself today and is more likely to promote world peace if it expressly and voluntarily waives the possession of atomic weapons of every kind."[2] Perhaps even more important was their statement that "none of the signatories would be willing to participate in the production, the testing or the use of atomic weapons in any way whatsoever." We see here one of those rare cases in which scientists expressly withheld their competencies from political-military functionalization.

Following the declaration of the "Göttingen 18," the Vereinigung Deutscher Wissenschaftler (Association of German Scientists) was founded. Today this association has about 300 members and sets itself the task of "awakening and deepening the awareness of those working in science for their responsibility regarding the consequences of their work on human society" and the task of "studying the problems for humanity which arise from the progressive development of science and technology." With its approximately 3,000 members, the Bund Demokratischer Wissenschaftler (Federation of Democratic Scientists) has similar objectives, as does the Gesellschaft für die Verantwortung in der Wissenschaft (Society for Responsibility in Science) with its approximately 200 members. These associations are to be strictly differentiated from academic and professional organizations. Generally, the latter refuse to make political statements, in part because such statements are prohibited by their articles of association. Conversely, associations such as the Association of German Scientists, the Federation of Democratic Scientists, and the Society for Responsibility in Science unite scientists on the basis of their individual, moral, political, or social involvement. As a result, these associations are politically powerless. Nevertheless, by means of public relations, contrastive research, and resolutions, they have had a certain political effect.

Scientists' resistance against the arms race has thus far been carried out by associations that unite individuals on the basis of their personal involvement. Therefore, the concept of responsibility has always played a special part in these associations. But can this type of organization and the concept of responsibility be the appropriate basis for resistance in the long run? For one thing, the collective character of scientific work cannot be translated into reality via the concept of responsibility. This concept presupposes that the individual, with the help of his conscience or God, can determine the responsible course of action. In reality, however, the will to act responsibly generally remains powerless precisely because the opportunities to act are not there. The division of labor in society continually undermines the possibility of actually putting responsibility into action. Therefore, the question is not so much one of responsibility as one of concrete political co-determination. In order to prevent the products of their research from being used in a manner they oppose, scientists must also attain the right to co-determine politically the uses and applications of science. Unfortunately, history has thus far shown that scientists gladly exempt themselves from science's context of application and do not wish even to take notice of the power they have by virtue of the significance of scientific knowledge in our society.

The Scientization of War and the Militarization of Science

Since World War II, the relations between science and the military complex have been both intensively and extensively augmented. Granted, one can speak of a long-standing affinity between scientific and military thinking. Not only exactness, clarity, and objectivity but also hierarchization apply to both. Since the age of absolutism there have been military academies, and the military corps has long been a source of institutes of technology (after mining schools). And since World War II the military sphere has been penetrated by scientific-technological thinking and thus also requires corresponding competencies—with respect not only to actual weapons systems but also to communications, reconnaissance/intelligence, and tactics. Officers today are often engineers or have some other academic training. Thus, the Bundeswehr (German Armed Forces) operates a number of engineering institutes (one in Darmstadt) and has two universities in the Federal Republic. Furthermore, because of the scientization of war, the greatest social resources can be mobilized for science. Because it is in this area that the greatest financial funding is available, because market rationality is disregarded in technological development, and because the latest and most sophisticated technology for the purposes of military superiority is always in demand, pioneering research projects are determined by military aspects in many natural-scientific and technical areas. Pioneering technology is thus also military technology.

To give an impression of the extent to which the military has subsumed natural science and technology, I shall cite some data from a work by Milton Leitenberg on the situation in the United States. According to Leitenberg, in 1973 more than 50 percent of R&D expenditure in the United States was related to military technology. The U.S. Army operated 44 research institutes, the Air Force 24, and the Navy 26. Moreover, 89 disciplines were sponsored by the Army and 60 by the Air Force. And in the 1960s, 1,000 doctoral dissertations were supported by military institutions. Leitenberg concludes: "There is no longer any distinction whatever between basic scientific research which may have military relevance and that which does not. This is not because science has changed but because the military 'requirements' have."³

According to UN estimates, about 400,000 scientists and engineers (i.e., about 40 percent of all scientific-technical personnel) are currently working in the area of weapons development worldwide. In the Federal Republic of Germany the situation is a little better than that in the United States because of Germany's promise not to develop nuclear weapons. Nevertheless, the official data presented in the Federal Research Report is frightening. According to this report, the expenditure for military R&D (DM 1.7 billion) in 1979 constituted 18.9 percent of the government's total expenditure on R&D. For a sense of the dimensions of this expenditure, consider the following comparison: The Max-Planck-Gesellschaft, Germany's most important institution for basic research, had for its 48 institutes in the same year a total budget of approximately DM 650 million; that is, it spent less than half of the official R&D expenditure for military purposes. For those of us in academia, military research is not quite so visible because it is increasingly being relocated to institutions outside the university. In Germany, the responsible bodies are the 16 Fraunhofer-Gesellschaft institutes, the Deutsche Forschungsgesellschaft für angewandte Naturwissenschaften (German Research Society for Applied Natural Science), the 31 institutes of Deutsche Forschungs- und Versuchsanstalt für Luft- und Raumfahrt (German Research and Test Institute for Aerospace) the Franco-German research institute Sain-Louis, and the Defense Ministry's own departmental research institutes.

War Research as an Institution

We have seen that the internal norms of the scientific community, such as its public character and universalism, have in specific cases enabled scientists to combat the functionalization of science for war. What determines scientists' behavior overall, however, is the fact that military science has become a permanent institution and that science has largely been institutionalized as military science. Sociology of science research has shown that the norms of the institutions employing scientists exert far more influence on the scientists' behavior than do the internal norms of the scientific community. These institutions also remove from individual scientists the burden of reflecting on the meaning of their actions. A mediocre, instrumental attitude toward work is all that one needs in order to pursue science.

On the basis of an interview with armaments researchers that appeared in the journal *Wechselwirkung,* I am compelled to ask how it is possible that people today continue to expend their vitality and brain power on the improvement of instruments of mass destruction. What kind of people are they? The answer is that they are entirely normal people, much like you and me. Virtually anyone studying natural science or engineering may, in the course of his or her professional career, be confronted with armaments research. As one armaments researcher put it in the above-mentioned interview, "Getting into this whole thing happened merely gradually."[4] In day-to-day work, these researchers suppress the fact that the technology in question is *armaments* technology: "The suppression functions very simply. When I am absorbed by a piece of work then I am not thinking about its usefulness, I am simply solving a problem."[5] This ability to suppress they trace back to the education they received at the university: "We are taught to view everything in naked technological terms and to problematize nothing, neither socio-political issues nor the effects of research." Again and again, they describe the work itself and the opportunities it provides as being especially attractive, even fascinating for the technologist. "No, it wasn't a question of salary. At best it was because one was told that there were interesting assignment areas to work on there. And then one would get to know qualified colleagues specialized in the various areas, and one could acquire additional qualifications because great things were expected from the equipment there. From today's somewhat detached perspective, I have to say that from this whole fascination at the beginning the only factor really remaining is the fact that in this area money plays a smaller part than in other areas. . . . And that of course makes possible a development with the most modern of methods. . . . That is, the fascination of always being at the frontiers of technology is still there."[6]

Is a Reversal of the Situation Possible?

Science for war: The problem is suppressed or pushed aside by an attitude that views science as just another job. When awareness of the problem does emerge, however, there is no lack of voices rejecting the involvement of science in military matters. The recommendations approved by the Pugwash-UNESCO symposium in Ajaccio (Corsica) in February 1982 thus state the following on this issue:

> The role of scientists is contrary to their traditional calling. The objectives of scientific endeavor should be a service to mankind, helping to better the fate of man and raise material and cultural standards. The basic unmet needs of a majority of the people in the world present a challenge great enough to warrant a huge and sustained effort by scientists. For an enormous effort of scientists to be instead directed towards wholesale destruction, to return to a state of primitive savagery among the survivors of a nuclear war, is an unforgivable perversion of science.[7]

In such a statement is evidenced a change of consciousness, one that no longer permits a legitimation of scientific work for military purposes. Yet the central problem is not thereby solved, as the enormous pressure of institutionalized military research has not been disposed of. Science for war has settled like a cancer in the organism of science; it has come to represent the whole to such an extent that many scientists believe that science would not survive an operation to have this cancer removed. Thus, the above-cited Pugwash-UNESCO statement emphasizes that there are still huge problems to be solved by science in the service of the well-being and peaceful existence of humanity. The catchword is "conversion": Not science for war, but science for peace.

Yet conversion is problematic, too—as scientists have a traditional aversion to praxis, to political conflict, to conflict as such. In this way the problem is pushed to the metalevel: Today there is more research on the concept of conversion than on concrete examples of conversion. The first and most necessary step would have to be a broad refusal campaign: Scientists—individually, collectively, and publicly—must refuse to make their competencies available for military purposes in the future. The Pugwash-UNESCO conference, too, is convinced of this necessity: "This world would be a much safer place if scientists in all countries would simply refuse to engage in military research."[8] The scientists assembled there were not, however, capable of calling on their colleagues to refuse; instead, they appealed to the conscience of the individual. This tender respect, this tolerance for those colleagues who might have different thoughts on the subject (or who act differently) conceals the fear of allowing conflict to break out within the institution of science itself. So real disarmament is impossible as long as scientists continue to work on the continuation of arms escalation: They have to recognize disarmament as being, above all, their own problem too.

Peace Research

One way science might redress the situation is to concentrate on the objective of "peace." On the basis of this insight, "peace research" has established itself in various parts of the world. An excellent example is the Stockholm International Peace Research Institute (SIPRI), which was founded in 1966. From the work of this institute came the simplest definition for what science for peace could mean: Its publications have been rated as being critical of the military, as antimilitary. Clarification of real military strength, discoveries of dangerous links between technology development and strategy, analysis and denunciation of the international arms trade, simulations of the course of possible wars—all such work falls within the realm of scientific competence and corresponds to the traditional enlightenment ethic of science. At SIPRI, science is science for peace in that it mobilizes a critical public sphere against militarism by providing cogent arguments and making the political world more transparent. Generally, however, it is only indirectly effective—namely, through the medium of the public sphere. Only in individual cases (such as the development of control instruments to secure the observance of arms control agreements) can science as such have a direct peace-securing effect.

It was therefore an important insight, adopted when the Deutsche Gesellschaft für Friedens- und Konfliktforschung (German Society for Peace and Conflict Research) was being founded in 1970, that science for peace must also relate to areas *prior to* military armament and conflicts. It must start with the causes of possible war—an insight that is still valid, even though the military complex has in the meantime become a largely independent and self-sufficient system. A research program has emerged that encompasses such issues as man's aggressiveness, the absence of an instinctual inhibition against killing the members of his own species, and the psychological and socioeconomic aspects of the emergence of conflicts. This largely analytical part of peace and conflict research is supplemented by one that relates to praxis. Those involved in the latter research, with the help of a peace pedagogics, to work toward the peaceful coexistence of human beings through socialization; also being developed are models of nonmilitary forms of conflict resolution and nonviolent resistance. Science can also be used in the de-escalation of existing conflicts—for instance, through destruction of projected enemy images. And, finally, it can be used to achieve an optimum interest satisfaction where there are conflicting interests—a kind of arithmetic justice.

That peace research exists at all gives rise to hope; moreover, its effect has been considerable, despite scant resources. (Investments in peace research are approximately one to ten thousand times smaller than investments in military research.) Science for war is, indeed, a perversion of science—because that whole enterprise called science should be in the service of peace.

The Asymmetry of War and Peace

The difficulties with war and peace are inversely related to the efforts that science has thus far invested in them. Not until recently—during a period when we can no longer comprehend war as a continuation of politics, when the abolition of war as such is humanity's task—have we begun to realize that the establishment of a permanent peace presents a far greater task for science than war has ever done. In other areas, too, people are starting to realize that the knowledge required for the maintenance and improvement of life is much more difficult to acquire than the knowledge needed for its destruction. Traceable back to Plato, and encompassing the instrumental use of science, is the idea that knowledge can be used for both good and evil. The doctor, notes Plato, can just as easily kill as cure patients with his knowledge. Yet the whole history of medicine has refuted this understanding. By committing a particular action, one can kill a person, straightforwardly and consciously, with only very partial knowledge. But the doctor's cure is today still dependent on nature's help. In the area of technology, the issue of war versus peace takes on similar implications: Within an extremely short period of time, less than ten years, knowledge about nuclear fission became the deadly knowledge of how to construct the bomb. But the decades that followed have not been sufficient to allow the program of a "peaceful use of nuclear energy" to mature. In contrast to the traditional ideology of science's value neutrality, it has become apparent that not every knowledge can be used just as easily for good as for evil. Knowledge that serves peace must be much more comprehensive; it must take into account relations far more complex. In fact, one could conclude that such knowledge is of a different type than the knowledge for war. This brings us back to the idea of a primordial harmony between military thinking and modern science.

The Idea of a Peaceful Science

In the chapter about technological utopias in *The Principle of Hope*, Ernst Bloch noted that technology and natural science still stood in nature like an army of occupation in enemy territory. Indeed, our natural science has developed epistemological methods that in no way relate to their object as a friend, an acquaintance, or even a relative; rather, they treat nature as something foreign to man, as something that threatens him and has to be mastered in order to serve human purposes. The success of the hard sciences has thus far been made possible by methods of opening and cutting, of isolating and controlling. Other sciences—particularly the human sciences—have been compelled to follow the same methods. Today it is clear that the successes of natural science and its compliant technology are extremely ambivalent. They were achievable only by assuming not only that other presuppositions could be neglected and side-effects could be regarded as insignificant but also that man's exchange with nature took place against the backdrop of an inexhaustible reservoir of natural resources and of nature's ever-reliable regenerative force. It is also clear that the endlessness of natural resources was an illusion and that nature's capacity to regenerate can be destroyed by human intervention. The conversion of scientific knowledge from the laboratory into large technological programs and processes that transform the countryside has shown that the side-effects of human measures vis-à-vis nature are not to be neglected. In medicine, too, it is becoming more apparent that the differentiation between intended main effect and materializing side-effect can no longer be maintained. These experiences permit one to assume that the knowledge type that has developed corresponds to an aggressive exchange with the object of knowledge. From a historical standpoint, this assumption is not surprising and seems even legitimate: Man first had to defend himself against an overwhelming nature, protect himself against its impact, and achieve a level of stability in the face of its fluctuations. He had to develop a knowledge that would enable him to master nature and have it in his service. Neglected at the same time, however, was a knowledge type that aims at cooperation, in the sense of man conceiving of himself as a part of a greater whole, as part of nature itself.

We must reckon not only with the fact that peace is a far greater task for man and his science than is war, but also with the fact that peace requires a different kind of science. In our century this insight has been appreciated by various authors such as E. Bloch, G. Bateson, and A. H. Maslow. To their minds, a science for peace would have to be understood as a knowledge type born of love for the object. No longer a knowledge of domination, it would have to be an orientational knowledge in order to guide man in greater contexts. It would have to be a knowledge of the whole, a system knowledge. And it would have to develop a technology that, following Bloch, would be designated "alliance technology"—a technology in which man understands nature as a partner.

Notes

This chapter has been translated into English by John Farrell.

1. Carl Schmitt, *Der Begriff des Politischen* (1932). Berlin: Duncker u. Humblot, 1963.
2. Fränkischer Kreis (ed.), *Dokumente zum Göttinger Manifest* (1957), p. 10.
3. M. Leitenberg, "The Dynamics of Military Technology Today," *International Social Sciences Journal*, vol. 25 (1973), p. 349.
4. *Wechselwirkung*, no. 9 (1981), p. 9.
5. Ibid., p. 11.
6. Ibid., p. 10.
7. *Pugwash Newsletter*, vol. 19, no. 4 (April 1982), p. 145.
8. Ibid., p. 145.

Bibliography

Acland-Hood, M. "Military Research and Development." In *SIPRI Yearbook 1983*. London: Taylor & Francis, 1983.

Autorenkollektiv, *Materialien zur Rüstungsforschung. Production und Imperialismus*. Darmstadt, n.d.

Bateson, G., *Steps to an Ecology of Mind*. London: Granada, 1973.

Bloch, E. *The Principle of Hope*, 3 vols. Cambridge, Mass.: MIT Press, forthcoming.

Bundesminister für Forschung und Technologie (ed.). *Bundesforschungsbericht VI*. Bonn, 1979.

Deutsche Gesellschaft für Friedens- und Konfliktforschung, *Dokumentation zur Tätigkeit der DGFK 1970–1983*. Bonn, 1983.

Fränkischer Kreis (ed.). "Dokumente zum Göttinger Manifest." *Schriftenreihe des Fränkischen Kreises*, 1 (1957).

Herbig, J. *Kettenreaktion. Das Drama der Atomphysiker*. Munich: Hanser, 1976.

Leitenberg, M. "The Dynamics of Military Technology Today," *International Social Science Journal*, vol. XXV (1973), pp. 336–357.

Maslow, A. H. *Die Psychologie der Wissenschaft*. Munich: Goldmann, 1977.

Myrdal, Alva. *The Game of Disarmament. How the United States and Russia Run the Arms Race*. New York: Pantheon Books, 1976.

Pugwash/UNESCO. *Ergebnisse und Empfehlungen des Pugwash/UNESCO Symposions*. Ajaccio (Corsica), February 19–23, 1982.

Rilling, R. "Militärische Forschung in der BRD." *Blätter für deutsche und internationale Politik* (1982), pp. 938–967.

Spiegel-Rösing, I., and D. de Solla Price. *Science, Technology and Society*, Chs. 10 and 12. London: Sage, 1977.

UNESCO. *World Directory of Peace Research Institutions*, 4th ed. United Nations Educational, 1981.

Wechselwirkungen, Heft 9. Topic: Wettdenken für den Krieg: Wer zuerst schießt, stirbt als zweiter (Competitive Thinking for War: Whoever shoots first dies second), 1981.

7

The Technological Civilization

In this chapter I propose to address the question of the meaning of science in our lives, both personal and social. I raise this question not in my capacity as a scientist, but as anyone might raise it insofar as his or her life is modified by the existence of science. I shall give some clarificatory examples: Our relation to our bodies is modified by the existence of effective drugs; the organization of social life is modified by social theories and the existence of computers; our nutrition is modified by chemistry, theories of nutrition, scientific agriculture, etc.; and our way of communicating with one another is modified by the existence of telephones and other communication techniques. From these examples, one may easily conclude that my question can also be formulated as the question of what life is like in the scientific and technological world. But this ushers in another question: are science and technology the same? I want to dispense with this question as quickly as possible since it has already yielded so much literature.[1] It has its background in the rivalries and struggles for status between scientists and engineers. It has been said that whereas science produces papers, technology produces things; that technology is the application of science. It has been said that science is oriented toward truth, technology toward utility; that science is about nature, technology about man-made things. As for me, science and technology are *one* human enterprise: the differences between them are differences of focus or emphasis. For example, technology is not only application of scientific laws, but it is modeling nature in accordance with social functions. But such modeling is what science already does. And while it is said that science forms the basis of technology, it can equally be said that technology constitutes the basis of science. In whatever way one might deal with this question, in our context it is science and technology as a whole which is relevant. The reason for that resides in the fact that the scientization of our lifeworld does not mean that we adopt a scientific attitude toward our lives, or that common sense is transformed into scientific thinking, but that, on the one hand, scientific products play an important role in our lives and that, on the other hand, certain actions are delegated to experts. For the layman, then, science and technology are a rather undifferentiated block.

This experience has a "fundamentum in re." There is a primordial unity of modern science and technology which stems from their origin in the Renaissance. There, for the first time, mechanics was accepted as a science, while nature was scrutinized under technical conditions (experimentation) and conceived in accordance with technical models (the clock, etc.). But in a sense the time was not ripe: The unity of science and technology had to wait until the nineteenth century to become effective.[2] That even today many authors insist on making this distinction we owe to the fact that Popper's conception of science as conjectural knowledge provides them with their understanding of science. Hence, the knowledge of engineers must be different from that of scientists. Indeed, the engineer cannot make much from conjectures. Fortunately, however, most aspects of scientific knowledge are not conjectural. It seems much more appropriate to make a comparison between science/technology and technics (technical apparatus and technical systems). It is true that one can, along with Karl Marx, call a machine "frozen science"; this is one form in which scientific knowledge is objectified. In this sense, even a great number of our social structures are "frozen social and political science," because they consist of social relations that have been reconstructed on the basis of scientific knowledge. The general trend of rationalization and scientization is to reconstruct our natural and social environment in accordance with scientific insights. But there remains a difference between a symbolic representation of knowledge and its incorporation in the structures of reality. Only in the first case may we say that knowledge is represented *as* knowledge. However, from the point of view of information theory there is no difference; both are representations of the same knowledge. And there is also no difference for the client, the layman: The natural as well as the social worlds may have become clearer and more controllable, but it is nevertheless the expert who is able to read scientific books as well as the structures of a scientifically reconstructed world.

To come back to our question, there is one sense in which the question "what is science?" must be answered by considering its effect on human life. Science has had such an important impact on human life that it cannot be sufficiently described as a type of knowledge or as the work of a certain social body—scientists—or as a particular relation to things. It must be seen as an epochal phenomenon, a phenomenon that has patterned a certain period of human history and is now going to transform the character of natural evolution.

There is no lack of good examples of theories that attempt to conceive of science epochally. The earliest of these date from the beginning of the nineteenth century. There is the idealist and romantic view that the emergence of science and philosophy is an evolutionary process in which nature becomes conscious of itself. There is Auguste Comte's theory of cultural evolution, according to which the scientific (positive) epoch follows the theological and metaphysical one. Furthermore, there is Max Weber's theory of rationalization describing a long-term process of European life which, finally, in our century, is supposed to penetrate the whole human world. Finally, there are theories trying to account for the particular reinforcement of the impact of science and technology on our lives over the course of the last few decades: the theory of the scientific-technical revolution, the theory of post-industrial society, and the theories viewing intellectuals as a new class.[3] I do not want to evaluate these theories, although they form the background to what I have to say. I would rather approach the transformation of human life directly. I wish to carry this out in two steps: anthropologically and sociologically. In other words, I should like to ask what scientization means to the individual and what it means to society.

Technical civilization is not yet fully realized, yet we are producing it. I say "producing" even though the conditions of life in technical civilization are not deliberately established but are rather side effects of processes of scientization pursued for other reasons. We are striving for an easier life, together with more effective work and faster communication. We want to improve human life, but we cause fundamental structural changes. Human life in technical civilization cannot be called straightforwardly "a better life," because it is a different way of life. These differences also can be (and are) seen by critics of our culture as drawbacks. It is in fact difficult to avoid their language, even if one only wants to talk about changes.

The Relation of Man to Himself

Let us begin with one of the changes that some may feel to be drastic because it touches upon a fundamental point of human self-understanding: namely, with the obsolescence of ethics. Ethics, we are told by the social sciences, fulfilled its function in achieving certain goals of society by regulating the behavior of individuals. It was important that the individuals were not aware of its social function. Its prescriptions were limited to being good or required. When at the beginning of the nineteenth century Malthus explicitly appealed to ethics in order to keep the growth of the population in check, he indicated by this very act that an epoch had come to an end. Today by technical means—namely, contraceptives—it is possible to aim in a straightforward way at this social goal, namely, a low birthrate. As a result, a great part of sexual morality becomes superfluous.[4] I do not want to claim that life in technical civilization will be a technically and no longer a morally regulated one. Morality will still exist, but it will be morality of a different kind. Lacking a social function, it will become a luxury closely related to aesthetics. This, by the way, is what ethics was in early Greek aristocratic societies.

The straightforward manner of bringing about desired effects conveyed by our example is quite characteristic of science and technology and can be generalized as operative in the relations men and women have to their bodies and souls. It is typical of an attitude fostered in us by science and technology, or better yet, by the products they offer, not to improve living conditions so as to reduce tuberculosis, but to vaccinate; not to count sheep to fall asleep, but to take a pill; not to look for adventure when we feel longing, but to switch on the television set. Scientific knowledge concerning how things are and the technical means for bringing about certain conditions provide us with an almost unavoidable temptation to take the direct route.

Let us consider in greater detail the human attitude toward body and soul characterizing technical civilization. Scientific knowledge of the human body is external knowledge. The decisive experience of the body is not self-experience but the experience of the other, i.e., the objective experience of a body which is not mine. Anatomy and physiology constitute the fundamental disciplines of medicine. Scientization of our respective relations to our bodies means that we adopt an attitude toward ourselves which, by virtue of its origin, models itself after the relation of a doctor to a patient. We very often treat our own body as if it were not our own body; we treat it as if we stood outside it. It is important to notice at this point that the utilization of scientific knowledge, and of medical and technical means, does not merely amount to introducing knowledge where there was ignorance, but to changing an attitude.

The situation is not as obvious when we examine the relation of man to his soul. Although "psychedelic drugs" are very often produced with the help of psychological and sociological knowledge, their consumption is still not grounded in the attitude described by "I have this or that psychic need, so I should satisfy it with this or that means." The most probable reason for this is that human desires are open to such a wide latitude of interpretation. We could say then that we go to see a film because we like cinema, and not because it fulfills a certain function within our psychic economy. But, in fact, it does.

To comment on this matter, we need to digress somewhat. But it is worth our while because it reveals another characteristic trait of technical civilization. To live in technical civilization means to live in an environment requiring an objective or cool and unemotional attitude. This environment is highly organized toward the establishment of security, i.e., toward the goal that nothing harmful should happen to the individual. As a consequence of these two characteristics, a great part of human emotional capacities and wants remains unfulfilled. On the whole, emotional life in technical civilization takes place in a detached, second world, a world of fictions. Technical civilization is therefore linked to the development of a huge imaginary world, which is steadily produced by the mass-media, and whose main function is to still emotional needs.

The detachment of emotional life from reality and the development of a merely fictional world began early in modern times and trickled down through the classes, with the spreading of an objective attitude to the world. It first culminated in extensive consumption of fiction by the eighteenth-century bourgeoisie, and a second time in movie consumption by the laboring class in the twentieth century.

A further change connected with technical civilization consists in the externalization of constraints. The far-reaching consequences and anthropological implications of this development can be seen only when it is considered against the background of Norbert Elias' research on the process of civilization.[5] Technical civilization is in some respects the reverse of what Elias described as the civilizing process. One of the hallmarks of this process is the internalization of restraints; what was once the effect of external restraints, namely peaceful and calculable behavior, became the effect of internal mechanisms, i.e., consciousness or the superego, when man became civilized. In hindsight, however, it becomes clear that Elias carried out his research or, better yet, was able to carry out his research, because he lived in a period when the wheel of history was beginning to turn around. Sexual morality was loosened, the threshold of shame was lowered, table manners were liberated, and educational practices became more permissive. Today, we can discern the reasons for these seemingly emancipatory developments: Inner constraints were no longer necessary because their function was overcome by outer ones. These external restraints are no longer exerted by masters, and do not consist in physical suppression. They can be called structural violence; they consist, in particular, of the constraints exerted by technical apparatus and technical infrastructure. To take an example, it is surprising to see the great importance attributed to punctuality by the older educational practices, and the draconian means by which it was enforced. It is surprising for this occurred at times when the need for punctuality was not so great. Today, on the other hand, when punctuality is important and time has become a fantastic system orchestrating the whole of social life, pedagogy no longer shows great respect for punctuality. The point is that punctuality nowadays must not be achieved by internal discipline, but is caused by external constraints—in particular, by the technical conditions of our life.

This decrease in inner constraints is related to—but is not identical with—a fundamental change in what counts as a mentally healthy, normal personality in technical civilization. One of the concepts of social psychology that had the greatest play during the past few decades was the concept of identity.[6] The reason for this is that symptoms of identity diffusion become more and more frequent. In other words, people felt it increasingly difficult to integrate the different parts of their life into *one* personality. Given a long-term perspective, this predicament might change. To have an integrated personality is a social requirement that loses its importance when the conditions of life, work, and communication change in a certain way. In recent times, many people—maybe the most "effective" persons—no longer felt the need to integrate their family role with their occupational role. The necessity to form an integrated personality will fade away when work stops being a "role" in itself—that is, work understood as a presentation of the whole person in a certain way—and becomes nothing but the execution of certain competences or merely the consumption of some portion of one's lifetime.

Social Relations

The obsolescence of ethics, the instrumentalization of the body, the detachment of emotional life, the externalization of constraints, and the compartmentalization of personal life are all changes which concern the structure of the personality in technical civilization. I shall now turn my attention to some characteristics of social relations, and thus move toward a consideration of the particular structures of society that are related to technical civilization. Firstly, the technical infrastructure of advanced societies redefines the social reality of an individual. In technical civilization, one exists, one is a person, provided that one has a connection or a terminal. I do not mean that this is or will be the only kind of social reality—the old ones, such as having a job, a vote, etc., still exist. But one should not underestimate the importance of technical integration. Today, a man without a phone is almost detached from social life; he is close to non-existence. This situation will become much more obvious when we have the new cable systems, i.e., when even more social activities—such as communicating and transmitting information, as well as ordering goods, working and even voting—will be mediated by these systems. It is no objection to but, rather, an additional argument in favor of my thesis that not every member of a family will have his or her own terminal: The same was true in traditional societies, where not everyone had the status of a person, and a family was considered one person.

This new definition of the social existence of a man or a woman is related to the fact that in technical civilization the bodily presence of a person is generally not required for him or her to act as a social entity.

It is difficult to get a clear idea of what this means because we can hardly say what bodily presence really is or was.[7] But one can at least give some hint. In traditional societies, to enter into a contract it was necessary to come together, to hand over personally what was sold, to break a stick in two pieces, etc. In modern societies, we retain the custom of signing contracts, the signature being a residue of bodily involvement. In technical civilization, social activities will be initiated by keys. We see a first example of this in the cards we use at cash machines or in automated gas stations. In some sense, this is not new. Marx said that man is the assemblage of his social relations. The new feature is that this assemblage is no longer anchored in the human body; the human body is becoming socially irrelevant. This need not mean that it vanishes from concern. On the contrary, its social irrelevance seems to be causing a new discovery of the human body. The more that social relations are reified, the more the human body becomes a vessel for individual revelation.

In technical civilization, the individual partakes of society as a terminal or a connection. We thus find a new type of social integration in technically developed societies. At the basis of this integration are technical infrastructures, that is, physical and mental supply systems, disposal systems, and traffic and communication systems. It is becoming ever more manifest that these systems form the basic structures of technological civilization. Not the single apparatus, but the systems within which such an apparatus alone makes sense and functions, continuously define what techniques really are. Not the bulb but the system of electrical supply, not the automobile but the system of cars, gas stations, and oil refineries, etc.—these are what pattern our way of life. Today these systems hold society together. As such, we have to invent a new term for social unity that is neither organic nor mechanical. According to Durkheim, mechanical unity or solidarity is unity through shared values; organic unity or solidarity is unity through mutual functional dependency, i.e., through the division of labor. The new unity of societies consists in an integration network. The term "network" leaves it open whether the systems are centralized or not. In fact, most of them are centralized, although they need not be. The important feature is that the individual becomes highly dependent upon these systems. There is scarcely anybody who can afford to be "disintegrated"; even terrorists can be traced insofar as they are consumers of electricity. Until now, these systems have not been used as a foundation of political integration. But this could easily be done—for example, by making personal numbers or other "keys" the precondition for participation in facilities of society.

Thus we have technical patterns for societies in our civilization. The technical infrastructures constitute a form of social skeleton. They must be seen as a new type of alienated superstructure alongside the state and the market (or capital). These superstructures are man-made but acquire a certain autonomy in that the individual becomes dependent on them and experiences them as endowed with their own dynamics, their own "intentions," etc., although he and his fellow men reproduce and develop this alien entity by their own actions. Much has been said about the autonomy of technology,[8] but the very essence of the phenomenon seems to be this: The autonomy of technology consists in the fact that we are literally living within technology; technical systems form the boundary conditions for our experiences and actions.

Technical Civilization and Culture

Thus far we have tried to say phenomenologically what the scientization and technical reconstruction of our world mean. One might be tempted to capture the different phenomena by saying that technology will become our new culture. In fact, it does assume some important functions once fulfilled by culture in traditional societies: the functions of orienting action, of defining the social order, of keeping society together. Furthermore, it can be observed that the progress of technical civilization dissolves traditional cultures on a worldwide scale. The way of living, the way of doing things and of resolving problems, the way of organizing communication and traffic, and the forms of production become detached from their original regional and cultural background and are reconstructed in a technical way. This gives the traveller the impression that the basic structures are nearly the same all over the world. But in fact technical civilization does not totally abolish or destroy traditional cultures, although it deprives them of some important functions. What happens can most easily be assessed when looking at our own historical experience. Culture, once a general framework of orientation, becomes a certain sector of life. Culture becomes a sector of politics, of production (cultural industry according to Horkheimer and Adorno),[9] as well as of private life. Culture belongs to our leisure time.

Thus it is illusory to say that technical civilization will become the culture of the new world. It is true that technical civilization has some characteristics which tend to be universalized, but technical civilization differs from culture. It engenders new dichotomies through which culture is not destroyed but set aside. This might seem to mean that culture becomes irrelevant. But it is important to get a clear idea of these dichotomies in order not to miss the possibilities inherent in the process. We have already pointed out that the phenomenon of the human body's social irrelevance includes the possibility of a rediscovery of the human body. We shall enumerate five of these separations or dichotomies:

1. Sexual intercourse is separated from generation.
2. Eating is separated from nutrition.
3. Fantasy is separated from perception.
4. Thinking is separated from calculation.
5. Movement is separated from traffic.

These examples show that the development of technical civilization is Janus-faced. While organizing life in a rigidly goal-oriented (*zweckrational*) way, it also opens fields for the development of new cultures. This two-facedness can be seen as analogous to the development of bourgeois society. The development of the bourgeois public, which put an end to the individual whim and subjected "everything" to law and public legitimation, on the other hand engendered bourgeois privacy and family intimacy. Thus technical civilization liberates in the twofold sense of "making irrelevant" and "opening-up fields of free activity, fantasy, and movement."[10] It may also set free personal relations and consciousness insofar as personal relations are no longer required to make a society and insofar as even certain types of thinking may be performed by machines.

This two-sidedness must be heeded if one is to define technical civilization. Jacques Ellul defined it by considering the kind of knowledge that becomes the dominant one.[11] For him, technical civilization is patterned in accordance with efficiency thinking. Lewis Mumford defined it in terms of the essential reality,[12] in our words, the superstructures of the technically reconstructed world or, in his, the mega-machine. Neither of them took into view the two-sidedness of the process, nor did any of the authors who considered the developing counter cultures, alternative cultures, esoteric and therapeutic practices, meditation, rituals, and so forth, as antagonistic to technical civilization. All failed to notice that the latter are the other side of the coin. Being aware of this, I shall define technical civilization as the disentanglement of purposeful action from the performance of life.

The same two-foldness must be kept in mind when one tries to evaluate the development of technical civilization. Is the tremendous effect of science and technology upon our world good or bad?

After centuries of enthusiasm about science and technology, hitherto interpreted as human progress, it is difficult today not to paint a gloomy picture and let development appear as a regress. In speaking about technical civilization and attempting to explicate what science and technology mean in our world, I wanted to make clear that they cause neither progress nor regress, but change. Our fate seems to be such that science and technology must be developed to a certain end, and thereafter it will be different to be a member of mankind.

Notes

I want to extend my gratitude to Pierre Adler and his collaborators for their excellent editing of this essay.

This chapter first appeared in the *Graduate Faculty Philosophy Journal* 12 (1987): 35–45. © 1987. Reprinted by permission.

[1.] For a survey of this literature, see I. Nordin, *Vad är teknik? Filosofiska funderingar kring teknikens struktur och dynamik* (Tema T Rapport 3, 1983, Universitetet i Linköping).

[2.] For the history of the unity and disunity of science/technology, see G. Böhme, W.v.D. Daele, and W. Krohn, "The Scientification of Technology," in *Finalization in Science,* W. Schafer, ed. (Boston: Reidel, 1983).

[3.] D. Bell, *The Coming of Post-Industrial Society* (New York: Basic Books, 1973); A. W. Gouldner, *The Future of Intellectuals and the Rise of the New Class* (New York: Macmillan, 1979).

[4.] For some further discussion, see P. Weingart, "Verwissenschaftlichung der Gesellschaft—Politisierung der Wissenschaft," *Zeitschrift für Soziologie,* 12 (1983), pp. 225–241.

[5.] N. Elias, *Über der Prozess der Zivilisation* (Frankfurt: Suhrkamp, 1976).

[6.] E. H. Erikson, *Identity and the Life Cycle* (New York: Norton, 1980).

[7.] I have given a discussion of bodily presence in my book *Anthropologie in pragmatischer Hinsicht* (Frankfurt: Suhrkamp, 1985).

[8.] For a survey, see L. Winner, *Autonomous Technology* (Cambridge, Mass.: MIT Press, 1977).

[9.] M. Horkheimer and Th. W. Adorno, *Dialectic of Enlightenment* (New York: Social Studies Association, 1944).

[10.] These possibilities may be missed through the renewed subsumption of the fields of free activity to technical rationality, as is the case with sports.

[11.] J. Ellul, *The Technological Society* (New York: Vintage Books, 1954).

[12.] L. Mumford, *Technics and Civilization* (New York/London: Harcourt Brace Jovanovich, 1962); L. Mumford, *The Myth of the Machine: The Pentagon of Power* (New York: Harcourt Brace Jovanovich, 1970).

8

The Knowledge-Structure of Society

It is not until recently that scholars began to look at the structure of society from the point of view of the social function of knowledge. Although, quite naturally, knowledge has always had a social function—e.g., as a demarcating principle in hermetic circles—knowledge had to wait until a few decades ago to be identified as a fundamental issue of sociological research. The reason for this most likely is that knowledge in the sense of science and technology and higher education became one of the most important parameters of advanced industrial civilizations. After having become part of the sociologist's consciousness, knowledge was then retroactively identified as a principle of domination in ancient water-societies and hierarchies. Applied to our type of society, the question is whether knowledge can provide the principle for social hierarchies, for the formation of a class structure, for the distribution of chances of social influence and personal life, and finally whether knowledge may serve as a principle of social cohesion and integration.

But before examining any of these questions, let me remind you of some of the facts which gave rise to attempts to reconceptualize what society is. One of the developments was the absolute size and rapid growth of science and technology itself. Science and technology had its beginning as a marginal enterprise of amateurs in the seventeenth century; but modern science, after the second world war, received a large proportion of the public budgets, and scientific-technological staff became a quantitatively important part of the labor force. D. de Solla Price, for example, calculated in 1963 that by the year 2000 every person will be a scientist if the exponential growth of science which was then observed would continue at the same pace.[1] Cause as well as effect of the growing importance of science and technology is the so-called explosion in the system of higher education in modern society. This is something we have experienced with particular force within the two last decades: Let me give you some figures taken from the West German census.[2] The proportion of the public budget spent on higher education has doubled within the last twenty years. Today it reached about 15 percent. The proportion of the age cohort which enters secondary schools went up from 26 to 49 percent. The proportion of the age cohort which went on to higher education tripled and in Germany now amounts to 19.5 percent. These figures become even more impressive if related to the size of the labor force. The extension of high school and university training together with the reduction of the life's working time led to a change in the relation of "learning" population to working population in 1960 to 1980 from 1:3 to 1:2. These figures alone are sufficient to show that our society has rapidly become a learning society as well as a producing society, that is, an "industrial society." It was Daniel Bell who in 1970 pointed to the fact that at present the institutions producing and reproducing knowledge, in one form or the other—that is, schools and universities—are comparable in size to the industrial complex. (In my home town, Darmstadt, there are two main concerns each employing more than 10,000 people: Ernst Merck chemical industries and the Technical University.) In short, science and education in our century became an important section of societies. Hence a particular "science policy" developed as well as an educational policy. But in recent times, we also observe grassroots movements engaged in the situation of schools and universities and others concerned with the dangerous aspects of certain technologies. The sociological awareness of the social and political importance of knowledge is only another effect of the impact of knowledge on the development of our society and our society's consciousness of this impact.

Let us now consider some of those sociological theories which have attempted to cope with the significance of science, technology, and higher education in our societies. The theories which I want to discuss are those of Radovan Richta et al., Daniel Bell, and Rudolf Bahro—and finally the theories which claim that the intellectuals are in the process of forming a new class.[3] I start with the ideas of Richta's Czech collective which have been published in the West under the title of "Society at the Crossroads." The ideas are known by the catch-phrase of the "Scientific Technical Revolution."[4] This theory is designed to comprehend the change of the social function of science within modern societies. This change takes place in two steps: Up to the end of the eighteenth century science has had the function of enlightenment, as a producer of *meaning*, in other words; in the following period it became a *productive* force. This is the first step. The second is that in our century science became an *immediate* productive force.

What does this mean? If science became a productive force in the nineteenth century it, as a result, does not belong to the superstructure any longer. The change from functioning as a producer or critic of world views to the function as a productive force means of course that science shifts its position in the structure of basis and superstructure. However, science has not yet become an immediate productive force. It is a productive force inasmuch as it is frozen into machinery. Inasmuch as it developed as "pure" science in the nineteenth century, it is not a productive force.

What is the reason that science began to play a role within the realm of production in the nineteenth century? My own answer is: The explanation is to be found in the degree of intellectual and material appropriation of nature in the nineteenth century. Earlier, science was not mature enough to be applied to problems of production; on the other hand, material appropriation of nature—in the sense of efficient control over boundary conditions, production of pure material, etc.—was not developed far enough to enable a realization of scientific results in dimensions relevant for production.

In our century—this is the claim of the Richta thesis—science has become an immediate productive force. "Immediacy" means, that, contrary to the situation in the nineteenth century, science may now be relevant for production, without being mediated by living (that is, corporeal) labor. Hence Richta and his group are talking about the coming abandonment of factory labor, about automation of production and the exterritorization of human labor from production into that of the preparation of production. The remainder of human labor will become, according to their theory, scientific and engineering labor.

The notion of the sciences as being an immediate productive force is not easily understood, even if the proposition about the coming abandonment of factory labor is accepted. It may be objected that the work of the engineer must be "realized"—that is, applied by the traditional labor of artisans or factory workers in producing machines for production. It may be asked: What does science actually produce as an immediate productive force? Let us first look more closely at the phenomena which concern Richta and his colleagues. E.g., they talk about the scientific penetration of labor and social practice. They hint at a process of analyzing, rationalizing, and generating data about labor and social practice: Realms of social life become subject to planning and control. Further, they refer to a structural change of labor in the sense of a change in the composition of the working class: It can be observed that there is an increase of higher qualification, in particular of the proportion of scientific-technical intelligence, a quantitative shift from the group of laborers employed within the section of production to the section of preparation, planning and regulation, and the services. Finally they call science a productive force because the leading branches of industry—that is, chemical and electric industries—are to a large extent science based. The consequence of all this is that the level of the productive forces today is largely defined by scientific-technological developments.

But let us go back again to our question as to why science is to be called an immediate productive force. What is the product of science? The answer must be: data and theories; let us say: data and programs. Hence science can be called an immediate productive force only in the event that data and programs as such are components of society. We have to face a situation in which the production of knowledge is immediate social production.

This, indeed, is the case. The reason for this is that a considerable part of the total work within our societies already takes place on the metalevel; it is second level production. Production to a large extent is not metabolism with nature any longer—that is, material appropriation of nature. Part of production presupposes that nature is already materially appropriated; it consists in rearranging appropriated nature according to certain programs. The laws which govern "secondary" production are not the laws of nature but the laws of social constructs: The consequence of this is that we have new types of disciplines, e.g., *Schaltalgebra,* computer science, and information theory. At the level of social practice we meet a comparable situation. The social sciences, which are seen as part of the immediate productive force by Richta et al., are sciences which have society as their subjects as it is in the state of being appropriated: operations research, cybernetics, theory of planning, and decision theory. Social sciences of this type presuppose that society is bureaucratically conditioned and prepared for data processing.

The result of this survey of the fundamental ideas of the theory of the scientific-technical revolution is this: It describes a phase of the societal development in which the production of knowledge is of fundamental importance for the reproduction of society. Reproduction of society means to an increasing degree reproduction of appropriated nature and of the self-appropriation of society. The production of data and programs is socially immediately productive, because it serves as such to reproduce the knowledge-structure of society.

Let us now turn to D. Bell's book, *Post-Industrial Society.* Bell tells us that we are leaving the age of industry. Accordingly, a structural change is taking place in our society: The production of goods will not be the characteristic of our society any longer, but the production of theoretical knowledge. Bell attributes a fundamental role to knowledge in the coming society; he designates it the "axial principle"—that is, the axis around which society revolves, the moving principle. Universities and research institutes are destined to become the central institutions of society; its main resource, scientifically trained man-power; its central problem, science policy and educational policy; its principle of stratification, skills and competences. The phenomena Bell refers to are not all that different from those identified by the Richta group—that is, alteration of the structure of labor, the shift within the working population toward services, the development of new technologies, the emergence of a new class, and the "educated people."

In order to understand the central point of Bell's theory, two questions have to be answered: What does he mean by "theoretical knowledge"? And what actually is the significance of knowledge as an axial principle in society?

It is obvious that Bell refers to science when he talks about theoretical knowledge—that is, a type of knowledge based on data, concerned with theory formation and empirical corroboration. The social change indicated by Bell has become obvious because the social significance which knowledge has always had has changed as a result of the fact that *scientific* knowledge is now required. Science-based industries are the most important ones; economic policy is performed with scientific means (Keynesianism); and "intellectual technologies" are developed by which problems of planning and evaluation become calculable.

What theoretical knowledge means as an axial principle can be made explicit if Daniel Bell's theory is compared with that of Karl Marx. According to Bell the axial principle in Marxian theory is the production of goods. Instead of the production of goods and its organization in the postindustrial society, the production of knowledge and its organization will be the most fundamental feature of society. That is, intellectual production will become more important than material production. The effect will be that the class structure of society, which had been dependent on the command of means of material production up to now, will increasingly be determined by the participation in and "ownership" of theoretical knowledge. This means that power in society which hitherto was based on the command of things will in the future be based on the command of information.

In order to explicate what the theory of "science as an immediate productive force" actually means, it was necessary to understand why the production of knowledge already is immediate social production. Toward this end, we had to emphasize the fact that in our societies a secondary structure on the basis of appropriated nature and society has been established. For science, quite naturally, never is productive in the same sense as the material productive forces that produce, say, machines or living corporeal labor. Daniel Bell's theory, to my mind, becomes understandable only when the function of knowledge for the structure of society has been theoretically explicated beforehand. Then it becomes possible to examine the effects of the shift toward a preponderance of theoretical knowledge. What is not sensible is to merely substitute knowledge production for the production of goods.

Let us now come to our third author—namely, Rudolf Bahro and his book *Die Alternative*.[5] He has already gone far with the conceptualization of social structures with respect to knowledge structures. For example, in contrast to Marxian theory, he insists on the fact that the alteration of property relations does not necessarily affect the structure of domination, that the accumulation and command of property very often was a consequence of domination, which had its fundament in the command of social knowledge. He further says that social stratification in the nations of "really existing socialism" is independent of property relations—which to him means that the Marxian concept of a social class has become obsolete. Social stratification, according to Bahro, is dependent on participation in knowledge and professional skills. A possible revolution for him is essentially a process of education; emancipation is the development of a personality whose participation in the life of society is mediated not so much by labor and consumption but by participation in the cultural life. The revolutionary power fostering this revolution, which is essentially a cultural revolution, is the so-called excess consciousness. As a possible agent, he proposes a new association of communists called the "collective intellectual."

The cultural revolution Bahro is talking about may be understood as a revolution in the superstructure of society. But what he really means is a shift of the center of society from the basis to the superstructure.

But requiring such a shift would be without sense, if the basis determines the superstructure. Hence we must overcome the schema of basis and superstructure if we want to make explicit the theoretical consequence of Bahro's procedure. Here again we meet the need for a theory of society centered on its knowledge-structure. As long as such a theory is missing we cannot understand why our society, thoroughly oriented toward the production of goods, produces an excess of consciousness, and why the old antagonism resulting from the different participation in the material resources of society is now going to be rivalled by another antagonism resulting from the different participation in intellectual resources of society.

Finally, we have to take two books into account which go one step further in conceiving of society on the basis of its knowledge structure: G. Konrád and I. Szelényi's *The Intellectuals on the Road to Class Power* and A. Gouldner's *The Future of Intellectuals and the Rise of the New Class*. Konrád and Szelényi investigate the role of intellectuals in the socialist states. They try to formulate a social theory for a society which has abandoned the private control over means of production. The authors describe these societies under the heading of "rational distribution"; that is, the surplus produced by the total society is distributed on the basis of rational planning. The intellectuals (e.g., social scientists, bureaucrats, planning engineers, and managers) therefore find themselves in the position of the old capital owners. Konrád and Szelényi point out that as a result of this structure a new class is emerging, access to which is regulated through certificates in higher education. This class has its own interests and is developing a growing antagonism toward the working class. Gouldner in a way has reproduced the analysis of Konrád and Szelényi for the Western hemisphere. The main difference is that the background is missing: In Western societies the private command of means of production is still largely working. But there is a similar development because the larger the companies grow the lower is the influence of private owners. This fact in earlier times already gave rise to the designation of the managers as the coming powergroup in our societies. Gouldner only claims the emergence of a new dominant class which is entering into competition with the existing class of capitalists. But he agrees with Konrád and Szelényi that the intellectuals are forming a class. This conclusion is based on Gouldner's assertion that knowledge is a type of capital. Capital according to Gouldner is to be defined as a resource which can be privately appropriated and on the basis of which one can earn one's living. It is true that knowledge cannot be privately appropriated in an exclusive way, but the access and the use can be regulated through privileges—say, degrees. As a consequence, the intellectuals develop class interests related to the access to and legitimated use of knowledge, and of the social benefits which can be derived therefrom. The basic thesis of the various theories of the new class therefore is: Knowledge is considered to be a social resource comparable to capital. Those people who participate in this resource develop group interests of their own, which have the tendency to bring them into an antagonism to other groups. There may even be a class domination, particularly when the interests of the intellectuals are in correspondence with the interests of administration.

Our survey of the theories, which try to cope with the fundamental changes in our society caused by the new position of knowledge, reveals that it is overdue to formulate a concept of society from the point of view of knowledge. In suggesting such a theory we may ask ourselves: How did we conceive of society up to the present? My own answer is: on the basis of property and labor. The original bourgeois society was a society of owners. It later became a "labouring society" (H. Arendt),[6] which in our day is going to be transformed into the "knowledge society." Property, labor, and knowledge, respectively, are the principles on the basis of which the individual can define his or her membership in society, by which the social position and the individual life chances are distributed, and which are responsible for the integration of the whole society.

In order to demonstrate the significance of knowledge for our societies in this way we first have to form a *sociological* concept of knowledge. Let us therefore differentiate between what is known, the content of knowledge, and "knowing" itself. What is it, that we know? I take some examples from the "Oxford Dictionary of Current English": "Every child knows that two and two make four. He knows a lot of English. Do you know how to play chess? I don't know whether he is here or not." These examples show that knowing is a relation to things and facts, but as well to laws and rules. In any case, I mean, knowing is some sort of participation: knowing things, facts, rules, is "appropriating" them in a way, including them into our field of orientation and competence. A very important point however is that knowledge can be objectified: that is, the intellectual appropriation of things etc. can be symbolically established, so that in the future in order to know it is no longer necessary to get into contact with the things themselves, but only with their symbolic representations. This is the social significance of language, writing, printing, data storage.

Let us designate the stock of knowledge which is established by symbolic representation "objectified knowledge." Please note that I do not restrict objectified knowledge to the stock of knowledge which is in some way materially put down, but include also what is canonized by social convention. The essential point is that most of what we today call knowledge and learning is knowing not about the things themselves but about objectified knowledge. Objectified knowledge is the stock of intellectually appropriated nature and society; it can also be called the "cultural resources" of society. We have defined "knowing" as participation in the things known. We can now state that knowing is grosso modo participation in the cultural resources of society. This formulation gives the adequate ground for explicating the social function of knowledge. Obviously the insider/outsider difference is dependent on the participation in a shared stock of knowledge. The fundamental example for this is given by a large community. But the internal differentiation of society can also be conceived on the ground of participation in different stocks of knowledge or provinces of meaning, as A. Schütz says.[7] The second point is that the interaction between members of a social group is mediated by a certain stock of objectified knowledge. What some sociological theorists have called symbolic interaction is exactly this type of interaction. There might be a more immediate type of interaction but it is hardly to be found in our societies. A third point we should mention is that objectified knowledge represents the accumulated knowledge of society about nature and about society itself. As a consequence, our life style, life chances, and social influence are to a large extent dependent on our ability to participate in "cultural resources." Finally, we have to account for the possibility that the access to certain stocks of knowledge may be restricted, that they, thus, may serve as a basis of hierarchies and domination.

These are rather sketchy remarks on the social meaning of knowledge; almost all can be derived from the sociology of knowledge of Alfred Schütz. But there is a significant difference in our "advanced societies," which after all made us ask for the knowledge structure of society. So let us ask what advanced societies are. You know that it is by no means obvious that the history of human society represents progress. On the contrary, it is a particular type of society which is to be characterized by this concept—namely, what we call modern or industrial society or technical civilization. Anthony Giddens, who has written a book about the *Advanced Societies,* means by that term "advanced industrial" societies.[8] But we have heard that others like Daniel Bell consider the progress now under way as something which is transforming the industrial society into something new, the post-industrial society. The fact is that the advancement of our societies does not affect the industrial sector alone, but affects traffic, communication, administration as well. This advancement can be called rationalization or modernization; it consists, in my opinion, in the progress of the intellectual appropriation of nature and society itself. There is an immense stock of objectified knowledge which mediates our relation to nature and to ourselves. Nature scarcely is experienced otherwise than as a human product or within human products, and the social relations of people one to another are mediated through highly sophisticated social, technical, and administrative regulations. In a traditional way we can state that our secondary nature is overgrowing the primary one.

More concretely, the material appropriation of nature means that nature as a whole is gradually transformed into a human product by superimposing to her a new, social structure. This structure in essence is objectified knowledge—namely, an explication and realization of what we know are the laws of nature extended by engineering design and construction. The same applies to society. The self-appropriation of society consists in the development of what we call culture; that is, the social life is organized according to rules. The first characterization which we can give for our situation is a mere quantitative one: The superstructure of society is so immense that the greater part of the overall social activity is not production but reproduction. Whereas in former times learning and socialization were only brief preparations for life, now they take up at least one-third of life.

This new prevalence of reproduction over production seems pleasant, as if the right relations had been reestablished. Actually it indicates only that our society in some way is an old society, which means that most of its production is repair.

The second consideration has much more to do with knowledge. Marx calls a machine frozen science. In the same way we can say that the superstructure of nature and society is *frozen knowledge*. Thus the main part of social production and reproduction is concerned with knowledge.

The third point is that within the structure of objectified knowledge that forms the superstructure of our society, scientific knowledge has acquired a dominant position. This is not only an historical fact, but has its basis in some of the significant characteristics of modern science as knowledge: E.g., modern science is impersonal, so it must not necessarily be transmitted in apprenticeship relations; it can be objectified. Modern science is universal; that is, it does not lose its validity when taken from its original context. Modern science follows the method of analysis/synthesis; thus is the adequate type of knowledge for "reconstructing" the world. And modern science is progressive; that is, it always takes a certain stock of knowledge for granted and its very activity is innovation on this foundation. In short, it provides an excellent medium for a developing social stock of knowledge. The more we go into details of typical traits of scientific knowledge, the more it becomes evident that the relation of science to our society—that is, a society which operates on and consists of a large stock of objectified knowledge—is not contingent. Scientific knowledge in our societies is not a type of knowledge among others, but the dominant type of knowledge, which in the final analysis is decisive. The meaning of knowledge within our societies can be summarized as follows:

1. In advanced societies the overall social activity is concerned with the reproduction of society. Because the superstructure of society is objectified knowledge, the institutions for cultural reproduction (education) are gaining an extent and an importance comparable to the institutions for material reproduction (agriculture and industry).
2. Objectified knowledge is a social resource equivalent to the material resources (capital). As a consequence, the social status and political influence become dependent on the degree of participation in knowledge.
3. Not all types of knowledge are equal. In advanced societies scientific knowledge is the dominant one, because it is intrinsically related to the developing superstructure of society. The differences between types of knowledge may cause social differences between

its carriers. Knowledge hierarchies in our societies tend to be replicated by social hierarchies.
4. There is a certain correspondence between administrative power and scientific knowledge. This again is a consequence of the fact that the superstructure of our society seen as objectified knowledge is essentially scientific. But it is this point which gives rise for a concluding consideration.

Conclusion

Summing up the new traits of advanced societies—the importance of objectified knowledge as a superstructure, the importance of the institutions of learning, the distribution of social and political chances according to the participation in knowledge, the production of knowledge as a main part of the overall social activity—we can advance the thesis that our society is going to be a knowledge society. In fact, it was this that Daniel Bell had in mind, when he claimed theoretical knowledge to be the axis of society, or Radovan Richta, when he talked about science becoming the leading productive force, Rudolf Bahro when he suggested the collective intellectual as the new revolutionary potential, or Konrád, Szelényi, and Alvin Gouldner when they considered the intellectuals to be the new dominating class. And it was knowledge as the new medium of sociality which gave rise to far-reaching hopes: for instance, that human labor would be to its main part creative intellectual labor, that social life would be much more a matter of participation in the cultural wealth of society than of material consumption. The new society was welcomed as the state of freedom. In fact, we are leaving the laboring society. If this was the state of necessity, does that mean that the coming knowledge society must be the state of freedom? What we observe is the contrary. Life's working time is shortening, the working part of the population is decreasing. But does this mean that people are set free for creativity and a new type of social relations? It is true that labor is losing its integrative power. It is true that people not working do not drop out of our societies; they are integrated by knowledge as the new principle of sociality. But they are integrated as *objects* of knowledge, not as its subjects. The more people are desintegrated from the laboring society, the more they become registered and controlled. Marginality by unemployment and homelessness—as well as by addiction, invalidity, prostitution, and sexual deviance—means registration. We observe that the self-appropriation of society today is going to mean a transformation of society into a huge data bank. The superstructure of society which we identified as objectified knowledge is becoming its total registration. So we are in danger that the new society which was welcomed as the knowledge society will appear in the form of *registration society*.

Notes

I am indebted to Professor N. Stehr, from Edmonton, Alberta, who read the first draft of this paper, for many helpful suggestions.

This chapter first appeared in G. Bergendal (ed.), *Knowledge Politics and the Traditions of Higher Education* (Stockholm: Almqvist and Wiksell, 1984). Reprinted by permission.

1. D. de Solla Price, *Little Science, Big Science* (New York: University of Columbia Press, 1963).

2. H. Meulemann, "Bildungsexpansion und Wandel der Bildungsvorstellungen zwischen 1958 und 1979: Eine Kohortenanalyse," in *Zt. f. Soziologie* 11 (1982), 227–253.

3. Radovan Richta (ed.), *Civilization at the Cross-Roads,* 3rd ed. (White Plains, N.Y.: International Arts and Sciences Press, 1969); D. Bell, *The Coming of Post-Industrial Society. A Venture in Social Forecasting* (New York: Basic Books, 1973); R. Bahro, *Die Alternative* (Frankfurt: EVA, 1977); and G. Konrád and I. Szelényi, *The Intellectuals on the Road to Class Power* (New York: Harcourt Brace Jovanovich, 1979). Part of the following analysis has been published in German, in Gernot Böhme, "Technologiekritik als gesellschaftlicher Konflikt," in Fr. Moser (Hrsg.), *Neue Funktionen von Wissenschaft und Technik in den 80er Jahren,* Wien: Verl. d. Wiss. Gesellschaften Österreichs (1981), pp. 52–76; and A. W. Gouldner, *The Future of Intellectuals and the Rise of the New Class* (New York: Macmillan 1979).

4. Arnold Buchholz, "Die Rolle der wissenschaftlich-technischen Revolution im Marxismus-Leninismus," 457–478, in Nico Stehr u. René König (eds.), *Wissenschaftssoziologie* (Opladen: Westdeutscher Verlag, 1975).

5. The English edition has been published by New Left Books.

6. H. Arendt, *Vita activa oder vom tätigen Leben* (München: Piper, 1981).

7. See, e.g., A. Schütz and Th. Luckmann, *The Structure of the Life-World* (London: Heinemann, 1974).

8. A. Giddens, *The Class Structure of the Advanced Societies* (London: Hutchinson, 1973).

9

An End to Progress?

The End of the Future?

The word "future" has had optimistic connotations in our culture ever since early modern times. When understanding ourselves as "in development," we expect a better, truer, and more essential being in the future. "Future" means the open frontier of the world, the direction in which man can evolve infinitely, the affluent field of opportunities. When thinking about the future, one feels liberated, fantasy is freed from the restrictions of the reality principle. Although we still experience this liberating impulse when thinking about the future, we are also disturbed; not only the principle of hope but also the principle of anxiety structure our ideas about the future. Utopias, depictions of a better world, are now balanced by "dystopias," visions of catastrophes to come.

Today we can talk about the future in a reasonable way only when—paradoxically—we dare to think this future might not even happen. Speculation about future developments is more or less explicitly modified by the presupposition that it not end in nuclear holocaust. However, this reference to the possibility of a dead-end is flat and irreflexive, for the notion of "peace as a precondition of the scientific-technological world"[1] is seen only as a border condition which does not modify our speculation about possible developments—for example, in physics, technology, or society. However, the unreflective continuation of partial developments may be exactly what precludes the precondition of this continuation—namely, that a third world war not occur. Hence, another type of apocalyptic thinking should intrude into our thoughts about the future: We should consider an end to history not in the sense of a finalization but in the sense of a termination of the accepted patterns of development.

This idea of termination causes some fear and mobilizes a defensive attitude. Although this effect may not be expected, it will appear plausible after some further comment. When the idea of an infinite universe arose in early modern times, it was welcomed by a very few persons, e.g., Jordano Bruno. For most people it meant metaphysical irritation since they were deprived of their accustomed horizon of a finite cosmos, and thus were submerged in a gulf of diffuse anxieties. Today, the situation is quite opposite: What frightens us is the idea that we might be forced to accept a finite world, i.e., a world without economic growth, scientific-technological innovation, or new resources. Obviously the idea of progress implies a tendency to run away—away from the present. This connection is well known when applied to economic growth: Economic growth relieves the conflict between labor and capital.

I would like to illustrate my claims that transformation in the pattern of development causes anxieties with an example that is particularly close to us scientists—namely, scientific development. When my colleagues and I from the Starnberg Max Plank Institute published our theory of the finalization of science,[2] we experienced some fierce political and emotional opposition.[3] This opposition was partly the result of a certain misunderstanding of our terminology. Whereas we intended "finalization" to mean goal orientation in scientific development, our opponents understood us to be announcing the end of science. Admittedly, they were not completely mistaken, for our theory actually implied that the number of fundamental problems within a particular scientific realm was finite. It is this idea which provokes resistance; it touches a nerve of scientific self-conception, according to which one is engaged in a process of unlimited progress. As to this self-conception of science, I would like to refer to a sentence of Popper's which was brought forward against our theory: "With each step taken forward, with each problem that we solve, we discover not only new and unsolved problems, but we discover also that where the ground upon which we stand was believed to be firm and certain, in truth everything is conceived of as uncertain and as vacillating."[4] This self-conception comprises some ideological elements because it serves to legitimate the need for scientists forever. Modern scientists conceive of themselves as researchers, and hence the prospect that some day nothing might be researchable causes anxiety. At the same time this point sheds some light on the paradoxical resistance to the ideal of truth in science. Although the search for truth is the understood motive of science, the possibility of arriving at truth is banned from the dominating theory of science. At the time of the finalization debate, we ironically termed this theory a "hydra-theory" of science—each question answered by science engenders some new unanswered questions. However, can we not imagine that the historical job of science might be done some day? We have to raise questions such as this one to sustain the irritation connected with them without anxiety. The fear of great catastrophes is the suppressed fear of changing our life patterns—one wants to continue as usual, and thus one runs the risk of the great explosion at the end.

The Concept of Progress

It is central to the pattern of development that determines our lives that human history implies progress. An example showing that this relation is conceived as something natural or even logical is the argument for increased armament. It is quite natural to modernize one's weapons. Clearly, one cannot be well equipped with the arms of yesterday for the war of tomorrow. However, the idea of progress is not a very old one in human history. The philosopher Karl Löwith eventually called progress the characteristic of merely a certain period in the history of humankind. Granted, even in antiquity there were some ideas about improvement and progress in different fields; still that does not mean that humankind was understood as evolving into an infinite horizon of open possibilities. For example, Plato tells a myth of progress in his dialogue *Politicos,* but this myth remains within the framework of an understanding of history as cyclical. Plato draws an ambivalent picture of human progress; it consists of a pair of divergent lines. On the one hand, science and technical competence become ever more efficient; on the other hand, there is a degradation of ethics and a loss of immediacy. This, according to Plato, must in the long run lead to unbearable conditions—so much so that one day God must interfere and turn back the wheel of history.

The interpretation of history as cyclical was fundamental until the Renaissance. This pattern of thinking included a certain concept of progress, but at that time progress always implied the advancement of humankind toward an ideal state that had been realized once before during the golden age. This meant that progress was never conceived of as infinite enhancement but, rather, as the attempt to return to an original state of completeness. According to the investigations of my colleague Wolf Krohn,[5] the modern idea of progress has two sources: One is the step-by-step improvement in artisanship and technology; the other is the humanistic idea of human development toward an ideal. Krohn denies that the theological notion of a history oriented toward salvation exerted any influence; but I think that it must be taken into account, for the concept of cyclical betterment held by the humanists had to be broken by a concept of linear time in order to form the modern idea of progress. However, another point of Krohn's theory seems important to me. He says that Francis Bacon and Descartes raised the understanding of progress as stemming from these sources to a metalevel; in other words, they made it the conscious principle of human history. According to Bacon and Descartes, what matters is that humankind achieves control over history by steadily striving for progress.

Summarizing this sketch of the emergence of progress, we can differentiate the following features. (1) The modern idea of progress implies an endless horizon in time. (2) Hence progress no longer means approaching a well-known ideal of completeness but finds its measure in the actual status quo. That is, progress becomes a dynamic principle—it is nothing but enhancement. (3) Thus the concept of progress contains a normative element; progress is the enhancement of what actually is. (4) The idea of progress becomes a principle of history. In other words, progress does not take place in particular dimensions but characterizes the development of humankind as a whole. (5) Artisanship and technology, and subsequently science, occupy leading positions in what constitutes progress for modernity.

I shall develop this last point somewhat further. Artisanship and technology initially provided the only clear and unquestionable examples of improvements; i.e., they offered the only examples of accumulation across time. The arts, rhetoric, and literature—comprising the proper field of humanism—experience losses as well as improvements and are thus characterized by their growing or shrinking distance from the ideal. Later, Bacon and Descartes designed science as a method of improving human conditions. In their eyes, science had to serve the improvement of human life through the mastery of nature. In addition, the scientific method had to provide a procedure for coping with social and human problems. This idea acquired some reality in the eighteenth century. Scientific method, understood as the procedure of analysis and synthesis, was effective as a principle of enlightenment. It caused a certain transparency, a liquification, and finally a liberalization of institutions. Keeping in mind this effect of science, we notice that our contemporary concept of progress is much more closely affiliated with the ideas of Bacon and Descartes than with those of Hegel and Marx, since the latter considered progress to be the penetration of human affairs by reason rather than scientific and technological improvement. When we raise the question whether progress as a pattern of human history may have come to an end, we are primarily concerned with science and technology: Do science and technology promise endless progress? The next question—immediately prompted by the preceding one—is whether this progress will be connected with human progress.

Progress in Science

In recent times, some voices—matched by supporting indications—have declared that scientific development may come to an end. To be sure, historical analysis tells us that such voices have been heard time and again, e.g., around the middle of the nineteenth century and again at the beginning of our own,[6] but these facts should not make us disregard the debate about a possible end of science. Let me first give a catalogue of relevant considerations.

The first may seem trivial because it deals with the quantitative growth of science. Already in the early sixties Derec de Solla Price[7] demonstrated that scientific growth is exponential, whatever measures of science you take (e.g., the number of articles published per year, the number of existing journals, the number of Ph.D. theses, the capacity of certain types of instruments). When these findings are compared with the possible resources in finance and manpower that are or may be available for science, it becomes clear that we have been in a critical phase for some time. Either scientific growth is running up against a wall (i.e., its over-complexity may no longer be manageable or its over-capacity may no longer be financeable) or scientific growth is slowing down drastically and will eventually resemble a logistic curve, which implies zero-growth in the long run. This has actually taken place during the last couple of years, and the transition turned out not to be catastrophic. Nevertheless, it was difficult in regard to the social and psychic situation of scientists.

However, the relevance of quantitative considerations may be questioned. What do they contribute to our question about whether science and technology may come to an end? One answer is that they account for the background of the "doomsday mood" connected with the end of growth. But we shall see later that quantitative considerations have wider implications.

A second consideration, the destruction of the belief in progress achieved by philosophy of science, should be mentioned. The leading philosophers of science originated this destruction of the belief in progress. Paradoxically, Popper is to be counted among the grave-diggers of the idea of progress. Did he not make growth of knowledge the central issue of philosophy of science? Science, according to Popper, is the endless process of improving hypotheses. Does this concept of science not include progress? Surely it does. However, since it neither acknowledges a footing in truth as a point of departure nor accepts a final arrival at truth, it comes as no surprise that in the aftermath of Popper's philosophy the image of science has been degraded to a mere network of hypotheses. When the generation of problems of science is finally considered as a merely internal process, as by Larry Laudan,[8] then science becomes nothing but a fantastic spider's web.

The other mode of destruction was initiated by Thomas Kuhn's book *The Structure of Scientific Revolutions.*[9] Kuhn's main attention was directed toward the fundamental revolutions in science. He has good evidence for his thesis that important scientific innovations do not simply enlarge the explanatory capacity of science but alter the world-view in general, which means that when new phenomena appear others disappear. Hence Kuhn himself stated that progress can no longer be considered as progress toward an end but only as progress away from an origin. This leads to a picture of science according to which each epoch is—as Ranke said—"equally distant to God." Thus science is on the same level as the arts. Nicholas Rescher drew precisely this conclusion:

> Today's major discoveries represent an overthrow of yesterday's: the big findings of science, it would appear, inevitably contradict its earlier big findings (in the absence of "saving qualifications"). Significant scientific progress is generally a matter of not adding further facts—on the order of filling in a crossword puzzle—but changing the framework itself.[10]

I would like to adduce a third consideration, namely, that of those voices coming from inside science, telling us that natural science in various domains—and perhaps as a whole—is nearing its final problems. Not surprisingly, very successful scientists repeatedly get the impression that the fundamental problems of their equally successful disciplines have been solved or will be solved very soon. Admittedly, this very often happens to be a mere confusion between biography and history. However, arguments are sometimes brought to bear. A case in point is Gunter Stent's book *The Coming of the Golden Age*.[11] According to Stent, an end to research may be expected in a discipline when the task is limited by its subject-matter. I quote from Stent's book:

> I think everyone will readily agree that there are *some* scientific disciplines which by reason of the phenomena to which they purport to address themselves, are *bounded*. Geography, for instance, is bounded because its goal of describing the features of the Earth is clearly limited.... And, as I hope to have shown in the preceding chapters, genetics is not only bounded, but its goal of understanding the mechanism of transmission of hereditary information *has*, in fact, been all but reached.[12]

This argument—if there is a definite task, it can be solved accordingly—is compelling but does not say much about science as a whole.

On the other hand, there are arguments that hold that the project of science as a whole may be limited. One such argument has been brought forward by von Weizsäcker.[13] Von Weizsäcker points out that the great branches of science such as physics, chemistry, and biology exhibit a tendency to grow together, and that they increasingly build upon one and the same ground. This historical fact is connected by von Weizsäcker with the Kantian idea that the true grounds of science are the conditions of the possibility of experience in general. This idea, in turn, is tied to the observation that a theory of elementary particles or fundamental

An End to Progress?

forces seems to become the basis of all scientific theories. What surpasses this idea is of yet only programmatic significance—namely, the project to develop a theory of elementary particles from general considerations about the necessary preconditions of experience. Von Weizsäcker's theory may be called a merely speculative one, but it is precisely its programmatic character which makes it superior to mere arguing about the eventual completion of science: It points to a way in which this completion can be achieved. I shall come back to this position later. But here I should add that the position in question is strongly opposed to the Popperian or Kuhnian understanding of science because it quite naturally implies that scientific questions can be given definitive answers.

The last part of this section will be devoted to Rescher's previously mentioned book, *Scientific Progress,* which may be considered to have set a standard for our present concern. Because of its brilliant argument and nearly comprehensive discussion of divergent positions, it provides an indispensable basis for further discussions of scientific progress.[14]

Rescher takes it for granted that economically science has to live with zero-growth. From this point of departure, he infers as a general trend a logarithmic decrease in the production of important scientific results:

> A simple but far-reaching *idée maîtresse* lies at the basis of these deliberations: the thought that if it requires (as over the past century or so it has) an exponentially increasing effort to maintain a relatively stable pace of scientific progress, then in a zero-growth era of constant effort science will enter a period of logarithmic deceleration.[15]

The decisive middle term of his argument apparently is the claim to "diminishing returns" in science, i.e., inversely the contention that continually increasing effort is necessary for obtaining comparably important results or breakthroughs. Empirically, this argument is neatly warranted by the fact that the expenditures for basic research in natural science have grown exponentially during the last few decades. It seems questionable, however, whether really important breakthrough and new phenomena can be obtained merely by attaining to new dimensions. This claim obviously rests on a qualitative estimation of results. But one may concede this to Rescher: He tries to find a more fundamental basis for his argument by the quasi-ontological supposition of levels within nature that follow logarithmic scales.

Summarizing Rescher's argument, it seems to be nothing but an oversophisticated explication of the qualitative impression many of us share—namely, that during the past decades steps toward solving fundamental problems could be made only with ever-growing expenditures of manpower, apparatus, and money.

Rescher's conclusion: There is no end to science, but only a deceleration of scientific progress. However, at this point an important difference emerges, namely, between synthetic and analytic problems. Rescher calls problems synthetic or power intensive if they can be overcome only with a compact effort. Examples provided are investigations undertaken in extremely small or extremely large dimensions, e.g., in particle physics and astrophysics. By contrast, analytical problems are problems of complexity that can be solved, as Rescher puts it, "in installments." The exponential expenditures necessary here must not be provided all at once, but could be dispensed over time. Analytical problems are to be found in biology and medicine. As for synthetic problems, Rescher does expect actual limits to science set by the limits of accessible support. But for analytic problems he looks forward to endless progress. Rescher thus thinks that a shift is taking place in science, a shift from fundamental problems to problems of complexity.[16]

Evaluating Rescher's argument within its own framework, one may come to the conclusion that it is self-defeating: on several occasions (e.g., on p. 53) he declares that the motive for writing his book was to stimulate society to maintain a steady effort in science. However, what actually follows from his argument is that really important results will become ever rarer. This insight will of necessity diminish the legitimation of the still considerable expenditures for science, and—what may be of greater consequence—it will discourage bright young people from becoming scientists. Consequently, an additional factor will contribute to the decrease in the rate of scientifically important findings.

However, we should also evaluate Rescher's argument with respect to considerations that did not come to his attention. As the preceding quotation revealed, Rescher is extremely Popperian: In his view, the only unquestionable dimension of progress in science lies in the instrumental domination of nature. He thus dwells upon a rather naïve concept of progress. He does not take the dialectic of the domination of nature into account, but he does adhere to the vision of man's capturing larger and larger dimensions and smaller and smaller particles, his entering the realms of shorter and shorter periods and deeper and deeper temperatures, etc. However, if there is no real progress toward truth, all scientific progress (i.e., progress qua domination of nature) must be considered rather ambivalent.

Rescher, on the one hand, does not know von Weizsäcker's argument, and on the other hand, in true Popperian fashion, is always oriented toward the esoteric frontiers of science. In other words, he does not care about what has been left behind—the classical theories, for example.[17] If we take these arguments and facts into consideration, Rescher's results appear in a different light. It is possible that in a time to come the basic questions of science will be answered—at least to the extent that they are questions concerning the fundamental building blocks of nature and inasmuch as these ultimate constituents can be experimentally identified and have practical value for humankind. Then we would have "closed theories" at our disposal, each of which would describe nature at a certain level of magnitude. Science—in full agreement with Rescher—would not have come to its end but would only develop in the direction of growing complexity. Then phenomena, natural processes with some practical impact on humankind as well as technically produced ones, would constitute the very field of research. Insofar as scientists would not set their hearts on investigating the colors of butterfly wings (Allison), technical problems (or problems engendered by technology) would provide the main subject-matter of science. Thus the question of the future of science shifts to the question of the future of techniques and technology.

An End to Progress by Science and Technology?

As far as technology is concerned there is no indication of an end or, on a certain level, even of a saturation. One reason is that in technology one essentially moves toward constructive complexity in the development of the world, a complexity for which there is no plausible limit. In addition the technical potential of scientific knowledge is by no means exhausted. On the contrary, some qualitative leaps may be expected from the solution of the remaining fundamental questions. There is no reasonable foresight in technology, only science fiction. To my mind there is only one possible end to technological growth—which may be illustrated by the story of Babel's tower. The technical transformation of the world is indeed akin to the building of the tower of Babel. The construction of the tower, the Bible tells us, was brought to a standstill by a confusion of languages. However, paintings such as Altdorfer's suggest another interpretation: namely, that one day the construction of the tower could not proceed any further because all of the manpower was already needed for repair. Our present second nature, the technical environment in which we live, is for the most part a heritage of our ancestors. This second nature in our time already consumes so much manpower and other resources for its reconstruction that one may imagine that some day in the future enlargement will no longer be possible. However—and at this point technical imagination again comes into play—it might be possible to develop self-reproducing technical systems. Biotechnology, the next clue to technology to come, might be a step in this direction.

If there are any doubts concerning progress in technology or in the whole complex of science and technology, these doubts do not concern the possibility of some further development; rather, they question whether technological progress has produced human progress. These doubts bear on the Baconian program, its legitimacy, and its feasibility. Although the actual enhancement of human life through science and technology must not be contested, doubts nevertheless arise when we observe that enhancements have meant deteriorations at the same time—that gains have been connected with losses. The fascinating improvement in control over nature at the same time has brought a frightening increase in man's power to destroy. Looking more closely, we must even say that the type of domination of nature which is provided by science and technology is more akin to destruction than conservation. Its manner of thinking—causal, linear, elementary—is much more capable of destroying a system than of keeping it alive.

The project of death is much easier to fulfill than the project of health. Let us take modern medicine as an example. The eminent achievements of scientific and technical medicine must not be denied. But they have not enhanced the average health of man. On the one hand, they have simply caused a quantitative shift among the different illnesses; on the other hand, they have contributed to the proliferation of exactly those diseases which they were able to treat, for instance, diabetes. The reduction in mother and pre-natal mortality did not improve the state of humankind as a whole. The well-being of some people has again been paid for by the unfortunate fate of others, who die of starvation and poverty. The extreme contemporary development of scientific and technological knowledge at the same time proves to be a huge process of unlearning as well as a devastation of non-scientific modes of knowledge. This has disabled the average man from being master of his life, and increased his dependence on experts. It is not even true that the work of the individual has been reduced by technical equipment. Indeed, what has been saved by technical means must be expended to provide for those very means. The car is a case in point:[18] What the men and women of our century save in labor-time they lose by the prolongation in traffic time. What may be saved in time for the preparation of meals at home must be paid for by a prolongation in shopping time. These are some examples of the ambivalence of technical achievements. Assuredly, one may wonder whether it is only the present "incomplete" state of our technology that reveals such ambivalence. Analyzing our contemporary technical thinking, we may discern deficiencies in it that suggest a different kind of technology. For example, the differentiation into the effects and side-effects of drugs is extremely short-sighted since it is biased by the appliers' interests. In reality—namely, at the level of the effects—there is no such difference. In addition to that, it is quite obvious that linear and elementary ways of thinking are not appropriate for complex systems. It is also clear that a technology which treats nature as a mere stockpile of raw materials fails to make use of and even destroys its reproductive forces. Consequently, the reproduction of the system in question will have to draw on human labor. Criticism of that sort may lead to a new technology. Today, however, we must say that the dominant technology no longer supports the hope for human progress.

The last statement might lead to our demanding an end to technological progress. In domains such as weapons technology, this in fact seems to be the appropriate demand. But I do not believe that the development of technology can be affected by moral demands. The motive forces for this development are to be found in economic conditions or, as is shown by the case of arms technology, in international power relations. Before claiming a new technology, one should envisage the truth that the technology that is ours and that determines our lives does not improve the human condition.

The question today is not how we can solve our problems by the application of technology, but how we can live in human dignity *under the conditions* of technology. Science and technology are no longer the means we can use to achieve this or that purpose, but are rather the boundary conditions of human life; they do not consist in individual insights or instrumentally usable things or apparatus, but form a basic pattern of our existence. Our way of living is not better than a pre-technical one; it is simply different. It implies other dangers and hopes, other modes of living and dying, other kinds of illness and health, other sorrows and joys.

Indeed, I believe that we have come to an end of progress, which means the end of an illusion concerning humanity's way toward betterment. The shattering of this illusion should not lead to lamentation, but should instead provide the occasion to consider what human life under technical conditions is and what particular opportunities it harbors.

Notes

This chapter was delivered at the Jerusalem Institute of the Van Leer Foundation, Jerusalem, in June of 1986.

This chapter first appeared in the *Graduate Faculty Philosophy Journal* 12 (1987): 237–250. © 1987. Reprinted by permission.

1. I owe this formulation to Carl Friedrich von Weizsäcker.
2. G. Böhme, W. v. d. Daele, and W. Krohn, "Finalization in Science," *Social Science Information,* vol. XV, 1976, pp. 307–330.
3. See the bibliography concerning the debate about finalization in the book edited by W. Schäfer, *Finalization in Science: The Social Orientation of Scientific Progress* (Dordrecht: Reidel, 1983).

4. Karl Popper, "Die Logik der Sozialwissenschaften," in T. W. Adorno et al. (eds.), *Der Positivismusstreit in der deutschen Soziologie* (Berlin: Luchterhand, 1971, third edition), p. 163.

5. W. Krohn, "Die 'Neue Wissenschaft' der Renaissance," in G. Böhme, W. v. d. Daele, and W. Krohn (eds.), *Experimentelle Philosophie* (Frankfurt: Suhrkamp, 1977).

6. See the documentation furnished by N. Rescher, *Scientific Progress* (Oxford: Basil Blackwell, 1978).

7. D. de Solla Price, *Little Science, Big Science* (New York: Columbia University Press, 1963).

8. L. Laudan, *Progress and Its Problems: Towards a Theory of Scientific Growth* (London: Routledge and Kegan Paul, 1977).

9. T. Kuhn, *The Structure of Scientific Revolutions* (1962), second ed., O. Neurath, R. Carnap, and C. Morris (eds.), *International Encyclopedia of Unified Science,* vol. II, 2 (London: University of Chicago Press, 1970).

10. Rescher, *op. cit.,* p. 48.

11. Gunther Stent, *The Coming of the Golden Age* (New York: Natural History, 1969).

12. *Ibid.,* p. 111f.

13. Carl Friedrich von Weizsäcker, *The Unity of Science* (New York: Farrar, Straus and Giroux, 1980), originally published in German in 1971.

14. Unfortunately, von Weizsäcker's positions do not seem to be known to Rescher.

15. Rescher, *op. cit.,* p. 2.

16. Incidentally, we stated the same conclusion in our original paper on the question of finalization: See note 2.

17. As for classical theories, see my article, "On the Possibility of 'Closed Theories,'" *Studies in the History and Philosophy of Science,* vol. 11, 1980, pp. 163–172.

18. This point is nicely illustrated in Ivan Illich's *Energy and Equity* (London: Calder and Boyars Ltd., 1974). In particular, see chapter VIII, which is a plea for the bicycle.

About the Book and Author

Gernot Böhme, a distinguished and original contributor to critical theory's philosophy of science project, sets out the main theses of this program in an important volume for science studies scholars. Stressing that science is a necessary aspect of advanced societies, Böhme explores the most fundamental questions about its social, political, and cultural roles in modern society.

In light of the mixed blessings of technical society, Böhme questions whether we can continue to regard the institution of science as the top of a hierarchy of knowledge or as a neutral means of progress, let alone as a benign force for good. Science and its future are too important to be left to the scientists; society, Böhme insists, must take control of its scientific future.

Gernot Böhme is professor of philosophy at the Technical University of Darmstadt. He is a contributor to *Finalization in Science,* edited by Wolf Schäfer, and coeditor with Nico Stehr of *The Knowledge Society.* Apart from the social research of science, Böhme's main fields of inquiry are the philosophy of nature, philosophical anthropology, and classical philosophy.

Index

Page numbers in italic indicate a definition of the term.

Subjects

Aristotle, 2
Atomic bomb, 65–70, 77. *See also* Nuclear power, peaceful use of

Bacon. *See* Baconian program
Baconian program, 1–17, 66, 112
  and socialism, 4. *See also* Scientific socialism, failure of
Bahro, Rudolf, 92, 96, 101
Bell, Daniel, 92, 94–95, 99, 101. *See also* Society, postindustrial
Body, instrumentalization of. *See* Case studies referred to, human body
Bureaucratization, 10, 51, 55

Case studies referred to
  classical hydrodynamics, 45–49
  human body, 10, 31–32, 84–87, 89, 90(n7). *See also* Case studies referred to, medicine
  medicine, 57, 78, 84, 112. *See also* Case studies referred to, human body
  midwifery, 10, 12, 58, *60–61*
  scientific obstetrics. *See* Case studies referred to, midwifery
Civilization, technological, 13, 81–90
Compartmentalization of personal life, 86
Comte, Auguste, 83
Cooperation of man and nature. *See* Man and nature, cooperation of
Culture of critical discourse, 14

Data generation. *See* Knowledge, acquisition of
Development. *See* Progress, scientific and human
Destruction potential, 9. *See also* Atomic bomb; Nuclear power, peaceful use of
Dialectic of progress. *See* Progress, dialectic of
Dichotomization of Western culture. *See* Western culture, dichotomization of
Disarmament. *See* Military disarmament
Disillusionment, 11

Ecology, 32
Education. *See* Knowledge, and education
Einstein's theory of relativity, 6, 22
Empathy. *See* Knowledge, through empathy
Ethics, 83–84, 86
  and the externalization of constraints, 85–86
  of science, 23, 68–71, 76, 113
  See also Sexual morality
Everyday life experience. *See* Life world
Experience. *See* Knowledge, scientific and non-scientific forms of
Expert dominance, 15, 51, 82

Fictionalization of life, 85, 89

Finalization in science. *See* Science, finalization in
Future. *See* Utopia and dystopia

Galileo, 50(n7)
Gouldner, 96–97, 101
Green revolution, 5

Humanity. *See* Science, social orientation of

Individual, ineffability of the, 29
Induction, 2

Kant, Immanuel, 21–22, 44–45, 48
Knowing. *See* Knowledge, and knowing
Knowledge, 2, 14, 26, 55–56, 57, 65, 98
  acquisition of, 35–38
  distribution of, 3
  and education, 13–14, 55, 91
  through empathy, 60–61
  hierarchies of, 10, 13, 23, 52, 55–56, 61, 100–101
  and knowing, 56–57, 58, 98
  and power, 55–56, 84, 91–92, 94, 97, 100–101. *See also* Nature, domination of; Society, knowledge-structure of
  scientific and non-scientific forms of, 8, 10, 13–15, 20, 23, 26, 30–35, 38, 49, 51–61, 78, 100, 112
  scientification of, 55–59, 82–84. *See also* Knowledge, scientific and non-scientific forms of
  traditions of. *See* Knowledge, scientific and non-scientific forms of
  See also Nature, knowledge of
Knowledge-based society. *See* Society, knowledge-structure of
Konrad and Szelényi, 96–97, 101
Kuhn, Thomas. *See* Science, normal and revolutionary

Life world, 10, 13, 38
  scientification of, 51, 81, 101. *See also* Society, scientification of
Logic, types of, 2

Man and nature, 9, 11, 21, 24, 30–36, 42
  cooperation of, 15
  See also Nature, domination of
Marxian theory, 95–96. *See also* Scientific socialism, failure of
Method. *See* Scientific method
Military disarmament, 8–9, 75. *See also* War and Peace, science for
Military research, 8, 73. *See also* Science, militarization of
Morality. *See* Ethics

Nature, 99
  destruction of. *See* Nature, domination of
  domination of, 3, 14–15, 23–24, 34, 60, 77–78, 99, 112–113
  knowledge of, 3, 14, 30–39, 42, 60, 99. *See also* Nature, laws of
  laws of, 21, 37–38, 45
  and man. *See* Man and nature
  theories of, 21
  See also Science, and nature
Navier-Stokes equations, 45–47
Newtonian mechanics, 6, 21–22, 37, 48
Norm, scientific. *See* Universalism
Nuclear power, peaceful use of, 5, 11, 77. *See also* Atomic bomb

Peace. *See* War and peace, science for
Peace research. *See* War and peace, science for
Physico-theology, 32
Plato, 21–22, 31
Popper, Karl, 21–26, 41–42, 48, 107, 109
Productive force. *See* Science, as productive force
Professionalization. *See* Science, social integration of

Progress
  cyclical concept of, 105
  dialectic of, 9–11, 14–15, 90
  fetishization and defetishization of, 13. See also Science, fetishization and defetishization of
  scientific and human, 1–2, 4, 8–9, 12, 15, 23, 25, 90, 105, *106,* 112–113
  See also Science, social orientation of; Scientific progress, limits of

Relativity. See Einstein's theory of relativity
Rescher, Nicholas, 109–111
Responsibility. See Ethics, of science
Revolution, scientific-technical. See Scientific revolution
Rictha, Radovan, 92–95, 101

Salomon's House, 3–4, 11
Science, 3, 8, 11, 13–14, *19,* 24, *26,* 29–30, *32–33, 38,* 39, 49, 65, 82
  alternatives of, 6
  applicability of, 4, 7–8, 14, 46, 48, 81
  and democracy, 51–53, 61
  development of. See Progress, scientific and human
  ethics of. See Ethics, of science
  fetishization and defetishization of, 4. See also Progress, fetishization and defetishization of
  finalization in, 6, 8, 11, 50(nn 5, 9), 104, 113(n3), 115
  institutionalization of, 8, 53–54, 87. See also Society, scientification of
  as an instrument of competition, 8
  limits of. See Scientific progress, limits of
  militarization of. See War, scientification of
  and nature, 9, 30, 34, 38–39
  normal and revolutionary, 6–7, 20, 25, 42, 107–108
  organization of, 3
  as productive force, 92–95
  promotion of, 5, 8
  relevance of, 6, 8, 14
  research. See Theory, of science
  social integration of, 3, 5, 24
  social orientation of, 3, 5, 7–9, 11–12, 14, 24, 65, 74, 106. See also Progress, scientific and human
  social studies of. See Theory, of science
  and technology, 8, *81*
  theory of. See Theory, of science
  universalism in. See Universalism
  usefulness of, 3–6, 9, 14, 24, 66
  and war. See War, scientification of
Science policy, 5, 8, 95
  priorities of, 6
Scientification of society. See Society, scientification of
Scientific and technological development, 1, 4–7, 9, 12, 15, 20, 33
  control of, 6
  and human condition. See Science, social orientation of
  phase model of, 6–7
  See also Progress, scientific and human
Scientific method, 2, 7, 13, 20–21, *26,* 32, 35–36, 100, 106
Scientific object
  formation of, 29–39
  unity of, 36–38
Scientific progress. See Progress, scientific and human
  limits of, 9, 29, 103–113
Scientific revolution, 31. See also Progress, scientific and human; Science, institutionalization of
Scientific socialism, failure of, 10, 97. See also Marxian theory
Scientific truth. See Truth, scientific
Scientization. See Scientification
Sexual morality, 84–85. See also Ethics
Social hierarchies, 91, 99
  disintegration of, 6, 54
  See also Knowledge, hierarchies of

Society
  knowledge-structure of, 91–102. *See also* Knowledge, and power
  organization of, 5, 10, 101
  postindustrial, 5, 83, 94, 99–100. *See also* Bell, Daniel
  scientification of, 10, 12, 15, 23, 25, 51–52, 56, 83. *See also* Knowledge, scientification of; Science, institutionalization of; War, scientification of
Szelényi. *See* Konrad and Szelényi

Technological civilization. *See* Civilization, technological
Technology
  ambivalence of, 89
  autonomy of, 88
  development of. *See* Progress, scientific and human
  See also Science, and technology
Theory, 44
  application of. *See* Science, applicability of
  closed, 6
  of science, 19–26, 104

Truth, 24
  consensus theory of, 43
  correspondence theory of, 43–45
  loss of, 1
  scientific, 21, 41–48, 49, 104

Unity, 87
Universalism, 52–53, 59, 66, 70, 100
Usefulness of science. *See* Science, usefulness of
Utopia and dystopia, 103

Validity, 43–47, 59

War and peace
  asymmetry of, 76–77
  science for, 9, 65–78
  See also War, scientification of
War, scientification of, 5, 8–9, 11, 23, 66, 69, 72–73. *See also* Society, scientification of; War and peace
Weber, Max, 83
Weizsäcker, Carl Friedrich von, 108–109
Western culture, dichotomization of, 13, 88–89
Wisdom, 15, 58

Proper Names

Ackernecht, E.H., 62(n5)
Acland-Hood, Mary, 17(n22), 79
Adenauer, Konrad, 70
Adler, Pierre, 90
Adorno, Theodor W., 17(n23), 29, 39(n2), 88, 90(n9), 114(n4)
Alembert, Jean Le Rond d', 31
Allison, Sam, 14, 70, 111
Altdorfer, Albrecht, 111
Anders, Günther, 13, 17(n28)
Andreae, Johann Valentin, 16(n5)
Arendt, Hanna, 97, 102(n6)
Aristotle, 2, 21, 30, 33, 39(n3)

Bacon, Francis, 1–15, 16(nn 2, 3, 6, 8), 25, 39(n6), 66, 105–106

Bahro, Rudolf, 55, 92, 96, 101, 102(n3)
Bateson, Gregory, 78
Bell, Daniel, 16(n15), 25, 27(n13), 90(n3), 92, 94–95, 99, 101, 102(n3)
Bergendal, G., 62(n2), 102
Bloch, Ernst, 77–78
Böhme, Hartmut, 17(n23), 39(n9), 50(n3)
Bohr, Niels, 70
Boyle, Robert, 32
Braun, Werner v., 66
Bruno, Jordano, 103
Buchholz, Arnold, 102(n4)
Bush, V., 27(n12)
Butta, R.E., 26(n6), 50(n3)

Index

Carnap, Rudolf, 27(n10), 114(n9)
Collins, H.M., 62(n12)
Comenius, Johann Amos, 1
Compton, Karl Taylor, 68
Cromwell, Oliver, 1

Daele, W.v.d., 15(nn 1, 17), 27(n11), 39(n7), 53, 62, 90(n2), 113(n2), 114(n5)
Descartes, René, 105–106.
Doorman, S.J., 17(n29)
Durkheim, Emile, 87

Einstein, Albert, 6, 22, 67
Elias, Norbert, 85, 90(n5)
Ellul, Jacques, 89, 90(n11)
Engelhardt, M.v., 62(nn 2, 10)
Erikson, E.H., 90(n6)

Fedotova, W., 16(n13)
Fermi, Enrico, 67–68
Feyerabend, Paul, 62
Fischer-Homberger, 62(n5)
Flügge, Siegfried, 50(n10), 67
Freud, Siegmund, 16(n16)

Galileo, 33, 50(n7), 66
Giddens, Anthony, 99, 102(n8)
Goethe, Johann Wolfgang von, 14, 17(n33)
Gouldner, A.W., 14, 17(nn 26, 31), 55, 90(n3), 96–97, 101, 102(n3)
Groth, Wilhelm, 67
Gurvitch, G., 63

Haber, Fritz, 66
Habermas, Jürgen, 50(n6)
Hahn, Otto, 67
Hanle, Wilhelm, 67
Harteck, Paul, 67
Heisenberg, Werner, 6–7, 17(n19), 67
Hegel, Georg Friedrich Wilhelm, 106
Helmholtz, Hermann v., 16(n11)
Herbig, Jost, 17(n32), 79
Hesse, H.A., 62(n7)
Holton, G., 26(n3)

Horkheimer, Max, 17(n23), 29, 39(n2), 88, 90(n9)
Husserl, Edmund, 49, 50(n11)

Illich, Ivan, 62(n2), 114(n18)

Joos, Georg, 67

Kant, Immanuel, 21–24, 26(n6), 35–37, 39(n11), 44–45, 48, 50(nn 3, 8), 108
Kistiakowski, George B., 68
König, René, 102(n4)
Konrad, G., 55, 96–97, 101, 102(n3)
Krohn, Wolf, 15(nn 1, 2, 17), 27(n11), 39(n7), 62, 90(n2), 105, 113(n2), 114(n5)
Kronick, B.A., 62(n4)
Kuhn, Thomas, 6, 7, 16(n18), 25, 27(n10), 42, 107–109, 114(n9)
Kutschmann, W., 39(n8), 49(n)

Laudan, Larry, 41, 50(n1), 107, 114(n8)
Lawrence, Ernest Orlando, 67–68
Leitenberg, Milton, 72, 78(n3)
Lenin, 16(n12)
Löwith, Karl, 105
Luckmann, Th., 102(n7)

Malthus, Thomas, 83
Manegold, K.H., 62(n6)
Marcuse, Herbert, 16(n16)
Marx, Karl, 82, 87, 95–96, 100, 106
Maslow, A.H., 19, 26(n1), 29, 39(n1), 78
Medawar, P.B., 26(n2)
Meja, V., 17(24)
Mendelsohn, E., 53
Merchant, Carolyn, 39(n4)
Merton, Robert K., 52–53
Meulemann, H., 102(n2)
Morris, C., 27(n10), 114(n9)
Moser, Fr., 102(n3)
Mulkay, M., 62(n3)
Mumford, Lewis, 89, 90(n12)

Myrdal, Alva, 79

Navier, C.L.M.H., 45–47
Neurath, O., 27(n10), 114(n9)
Newton, Sir Isaac, 6, 22, 32, 37, 47–48
Nordin, I., 90(n1)
Novotny, H., 17(n25)

Oppenheimer, J. Robert, 68
Oswatitsch, K., 50(n10)

Parsons, Talcott, 62(nn 7, 11)
Paulsen, F., 62(n8)
Pericles, 21
Plato, 21–23, 26(n5), 31, 77, 105
Polanyi, M., 26(n4), 62(n12)
Popper, Sir Karl, 21–25, 26(n7), 27(n8), 41–42, 48, 82, 104, 107, 109–110, 114(n4)
Puntel, L.B., 50(n2)

Ranke, Leopold von, 108
Reichenbach, Hans, 27(n9)
Rescher, Nicholas, 108–109, 114(nn 6, 10, 15)
Richta, Radovan, 16(n14), 27(n14), 92–95, 101
Rilling, R., 17(n21), 79
Roosevelt, Franklin Delano, 67

Schäfer, Wolf, 17(n20), 50(nn 5, 9), 90(n2), 113(n3), 115
Schilpp, P.A., 26(n7)
Schmitt, Carl, 66, 78(n1)

Schnabel, F., 62(n8)
Schütz, Alfred, 98–99, 102(n7)
Skirbekk, G., 50(n2)
Sncw, Sir Charles Percy, 13, 17(n30)
Socrates, 21
Solla-Price, Derec de, 79, 91, 102(n1), 107, 114(n7)
Spiegel-Rösing, I., 79
Stehr, Nico, 17(n24), 26(n5), 102, 102(n4), 115
Stent, Gunter, 108, 114(nn 11, 12)
Stokes, Sir George Gabriel, 45–47
Straßmann, Fritz, 67
Szélenyi, I., 55, 96–97, 101, 102(n3)
Szilard, Leo, 67–68, 70

Thales, 60
Teller, Edward, 67
Törnebohm, H., 49(n)

Velicovsky, 62(n3)
Virchow, Rudolf, 51, 62(n1)

Weber, Max, 2, 16(n9), 24
Webster, C., 16(n4), 53
Weingart, P., 90(n4)
Weizsäcker, Carl Friedrich von, 6, 7, 17(n19), 67, 108–110, 113(n1), 114(nn 13, 14)
Wigner, Eugene Paul, 67
Winner, L., 90(n8)
Wittgenstein, Ludwig, 48

Znaniecki, F., 16(n7), 62(n13)